PUBLICATIONS OF
THE WELLCOME HISTORICAL MEDICAL MUSEUM

No. I

A PRELUDE

TO MODERN SCIENCE

A PRELUDE
TO MODERN SCIENCE

BEING

A DISCUSSION OF THE HISTORY, SOURCES AND
CIRCUMSTANCES OF THE
'TABULAE ANATOMICAE SEX' OF VESALIUS

BY

CHARLES SINGER AND C. RABIN

Published for

THE WELLCOME HISTORICAL MEDICAL MUSEUM

at the

UNIVERSITY PRESS
CAMBRIDGE

1946

CAMBRIDGE UNIVERSITY PRESS
Cambridge, New York, Melbourne, Madrid, Cape Town,
Singapore, São Paulo, Delhi, Mexico City

Cambridge University Press
The Edinburgh Building, Cambridge CB2 8RU, UK

Published in the United States of America by Cambridge University Press, New York

www.cambridge.org
Information on this title: www.cambridge.org/9781107600690

© Cambridge University Press 1946

First published 1946
First paperback edition 2012

A catalogue record for this publication is available from the British Library

ISBN 978-1-107-60069-0 Paperback

♣

Dedicated with respect and affection to
D'arcy Wentworth Thompson
and
Charles Scott Sherrington

PREFACE

ANATOMY was the first observational discipline to be studied on modern scientific lines. The *Tabulae Sex* of Vesalius of 1538 is almost ideal material for the investigation of its beginnings and is thus itself a prelude to modern science. We seek here to display some of the forces which converge to make Vesalius a figure of such scientific importance, some of the currents of thought which enter into the *Tabulae*, some of the relationships of the *Tabulae* to its predecessors and to contemporary activities, and some of those elements in the *Tabulae* which foreshadow the masterpiece of Vesalius, his immortal *Fabrica* of 1543.

We have had much help in our work and it is a pleasure to acknowledge our various debts. The Wellcome Trustees have paid part of the expense of the research. We desire to thank the Wellcome Foundation and the Wellcome Historical Medical Museum for publishing this work, and for defraying the whole of the cost of printing and publication. The great resources of the Library of the Museum have been at our disposal through the kindness of its Director, Dr. Ashworth Underwood ; to him and to the Librarian, Mr. W. J. Bishop, we are indebted for much assistance. We have had similar generous help from Mr. Home of the Royal Society of Medicine and Dr. W. R. Cunningham of the Library of the University of Glasgow. Dr. F. P. Reagan, Reader in Anatomy in the University of Birmingham, has drawn figures 21, 24, 34 and 35, has re-drawn figure 23, and has made many helpful suggestions ; figure 47 is also his work. Miss D. M. Barber, Artist to the Central Middlesex County Hospital, has kindly drawn figures 44, 45, 53 and 54 from the objects. Material for dissection has been provided by the Zoological Society of London and by Professor Legros Clark, F.R.S., of Oxford. Philological and syntactical puzzles have been solved by Dr. H. A. R. Gibb, Laudian Professor of Arabic at Oxford, and by Dr. E. Fraenkel, F.B.A., Corpus Christi Professor of Latin at Oxford.

Help and advice on other special points have been given by Professor A. J. E. Cave of the Royal College of Surgeons of England (now Professor of Anatomy at St. Bartholomew's Hospital), Professor F. J. Cole, F.R.S. of Reading, Dr. Kenneth Franklin and Dr. Cecil Roth of Oxford, Mr. A. M. Hind, O.B.E. of the British Museum, Dr. Max H. Fisch, Head of the Rare Book Division of the U.S. Army Medical Library, Dr. John Humphrey of University College Hospital, London,

a

PREFACE

Mr. C. M. Ivins of the Metropolitan Museum of Art, New York, Professor Saxl, F.B.A. of the Warburg Institute, Dr. G. H. Streeter of Boston, and Dr. Josiah C. Trent of the University of Michigan Hospital. To these too we tender our grateful thanks.

The figures of the *Tabulae* at the end of the volume are reduced to about half diameter from a copy of the Stirling-Maxwell facsimile in the Wellcome Library. It seemed needless to reprint the Latin text since it is quite legible even in this reduced form. The figures from the *Fabrica* and *Epitome* in the body of the volume have been taken from the facsimiles in the *Icones*. All references to the text of the *Fabrica* are to the edition of 1543 unless otherwise stated.

We have translated the entire text of the *Tabulae* except for the emptily repetitive compliments in its opening paragraphs. It was found impossible to render these into English, a language happily unable to sustain the multi-involuted entanglements of renaissance sycophantic logomachy. The anatomical part of the text contains so many difficulties for the modern student that it has seemed best to provide a running commentary, paragraph by paragraph, rather than relegate the exposition to the end of the work. The notes that make up this commentary have, however, been numbered consecutively for easy reference by readers of our Introductory Sections.

Much of the discussion turns on the passage of anatomical terms from Arabic, Hebrew and Greek into Latin or modern vernaculars. It has seemed advisable, therefore, to avoid alphabets unfamiliar to ordinary readers. In all cases we have transliterated Arabic, Hebrew and Greek words in Latin letters, adopting recognised conventions. Very little is lost and much is gained by this practice, which should, in our opinion, be more widely adopted. We trust that it may help anatomists to trace the history of some of the terms that they employ. We hope, too, that the Summary of the work that follows this Preface may be useful to those who do not pursue our argument in detail.

August 1946.

C. S.
C. R.

SUMMARY AND TABLE OF CONTENTS

I. Character and Purpose of the 'Tabulae'.

In the *Tabulae* of 1538 renaissance classical scholarship, based on newly recovered Galenic texts, is grafted on medieval tradition. The *Fabrica* of 1543 is essentially a product of modern scientific research though it is set out in the renaissance humanist manner. What induced this change of outlook in the intervening years?

Renaissance art influenced anatomy late. Many sixteenth-century anatomists opposed the graphic method as contrary to Galenic tradition. The *Tabulae* contains the first attempt to represent the vascular system by figures. Graphic anatomy soon destroyed astrological medicine.

It was printed from wood-blocks. The first three tabulae are diagrams illustrating Galenic physiology. They were drawn by Vesalius and show him as a competent draughtsman. The last three, representing the skeleton of a rickety male of 18, were by Calcar and are artistically below what might be expected from a pupil of Titian.

II. Vesalius and the Schools of Louvain and Paris.

In 1528, at fourteen, he entered the old-fashioned *Pedagogium Castre* at Louvain. In 1531 he transferred to the humanistic *Collegium Trilingue* where he learnt some Greek and became infected with the spirit of a *trilinguis homo*. He left for Paris in 1533.

Dissections of medieval type began at Paris c. 1493. Latin versions of Galen's anatomical works began to appear in 1528 with the *De usu partium*, etc. Guenther arrived in 1527 and translated Galen's *De anatomicis administrationibus* in 1531. His teaching, represented by his *Institutiones anatomicae* of 1536, introduced many modern anatomical terms. Sylvius began teaching in 1531. His teaching, seen best in his *In Hippocratis et Galeni physiologiae partem anatomicam Isagoge* (1542), introduced the modern nomenclature of muscles and the modern method of enumerating vessels, nerves, etc.

He was ten years older than Vesalius but took his medical degree six years later, had a better classical training, and was anatomising before Vesalius reached Paris. Almost all of his *De dissectione partium corporis humani* (1545) was in print before the *Tabulae*. His anatomy did not influence Vesalius nor was it influenced by Vesalius.

He was not the pioneer of dissection at Paris. His dissections there were made to verify the texts of Galen. He did such dissections as were needed for Guenther's *Institutiones* (1536).

He dissected his first female body in 1536 at Louvain and found a *corpus luteum.* Late in 1536 he prepared his first complete skeleton. Early in 1537 he made the first 'anatomy' at Louvain for eighteen years. It involved him in minor controversy. He printed at Louvain in 1537 his *Paraphrasis in nonum librum Rhazae ad Almansorem* at Louvain. It is a pretentious work and contains the *suggestio falsi* that he could read Arabic.

When he reached Italy he knew only the ordinary Galenic anatomical texts. The editions accessible to him are enumerated.

III. Italian Predecessors of Vesalius.

Valla of Venice was the first to use Pollux (*c.* 1490). Benedictus of Padua, who had a special building for dissection, was the first after Mundinus to write a book on anatomy (1502). Benivieni (*d.* 1502) of Florence left case-notes on autopsies. Gerbi (*d.* 1506) of Padua and Bologna, a 'scholastic' anatomist, improved the accounts of the abdominal viscera. Achillini (*d.* 1512) of Padua and Bologna, an argumentative Aristotelian, made several anatomical discoveries but hardly influenced anatomy.

The first medical work influenced by renaissance art is the Latin *Fasciculus medicinae* Venice, 1491, the first showing an internal organ drawn from the object is the Italian *Fasciculo di medicina*, Venice, 1493. The first illustrated anatomies are those of Berengar of Carpi, printed at Bologna between 1519 and 1523. He dissected many bodies and improved the descriptions of many organs.

His *Anatomy* (Venice, 1536) is based on many dissections, including some full-time foetuses. He began dissecting long before Vesalius who must have known his work. He has the first description of the prostate and good accounts of liver, kidney and seminal vessels. He rejected the 'third ventricle' of the heart but accepted the *rete mirabile.* He recognised anatomical differences between apes and men.

IV. Galenic Physiology and its Latin Presentation.

The world-spirit, *pneuma* or *anima*, the basis of his system, is inconsistent with orthodox theology.

Essential elements are (*a*) the elaboration of 'natural spirit' in the vessels of the liver, (*b*) of 'vital spirit' in the pores of the cardiac septum, and (*c*) of 'animal spirit' in the *rete mirabile.* To these must be added the four 'temperaments' or 'complexions' and the four 'humours'.

V. Certain Anatomical Elements in the 'Tabulae'.

VI. Renaissance Anatomical Vocabulary.

VII. Semitic Elements in the 'Tabulae'.

SUMMARY AND TABLE OF CONTENTS

PLATES

Facsimile of the 'Tabulae'　　　　*at end*

VESALIUS

TABULAE ANATOMICAE SEX

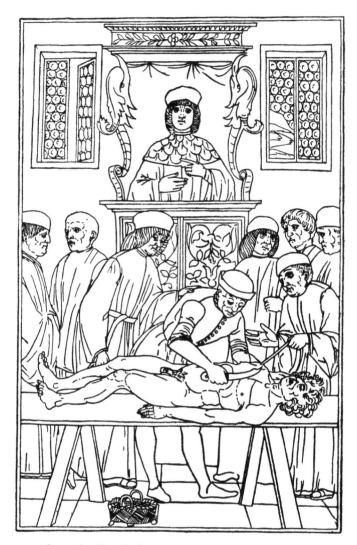

FIG. 1.—Dissection scene from the *Fasciculo di medicina*, Venice, 1493. It represents an academic anatomy at Padua University. The professor, in academic dress, reads his text from his 'chair' while a menial performs the dissection, directed by a 'demonstrator' in cap and gown. Reduced to ½ diam. from the original, which is stencilled in four colours.

VENETIIS ANNO .D. M.CCCCC.XXXV.

FIG. 2.—A dissection scene from an edition of Berengar's *Isagogae*, Venice, 1535. The conception of an anatomical demonstration is still that which had prevailed in the fifteenth century. Compare with Fig. 1.

I. Character and Purpose of the 'Tabulae'.

§ 1. *The Vesalian Problem.*

The work of Vesalius is of high significance for the development of philosophy and of science for it heralds the modern spirit of research. His importance extends far beyond the realm of Medicine, and no figure in its history since antiquity has elicited so extensive a literature. It is appropriate to begin with a brief attempt to explain why this should be.

The name of Vesalius must always be associated with a masterpiece, the *De humani corporis fabrica* and its supplement, oddly misnamed as its *Epitome*. Both were issued at Basel in 1543. The two books can for many purposes be treated as one. The combined work is the first great achievement of modern observational science. It opens a new scene as with the quick rise of a curtain, for it is suddenly, essentially and brilliantly modern—modern in appearance, modern in outlook, modern in method, modern in its art and in its technique. And modern, too, in what it omits no less than what it includes. It is a startling apparition in the very midst of that imitative world of the revival of classical learning which still pervades its language. To appreciate at all the magnificence of the great adventure of the *Fabrica* we must know something of its starting-point. The process of obtaining the mass of knowledge that it contains was a true voyage of discovery. The work that we here seek to elucidate is the port of embarkation.

All who have looked into the *Fabrica* and its history are aware of the existence of a very complex series of questions, summarised as 'the Vesalian problem'. The basic mystery is the abrupt intrusion into a non-scientific renaissance society, the intellectual interests of which were centred on the ancient classics, of an immense and highly finished monograph in a quite new manner. Nothing of the sort or class had been printed before. Leonardo's work was its only comparable predecessor, and Leonardo never published anything and never finished anything. He could not have influenced Vesalius for the two men were too widely separated in space, time and outlook.

The *Fabrica* introduced a new method of representation to the printed book. Who drew and cut these remarkable figures ? The more traditional skeletons of the *Tabulae*, which we shall presently discuss, are certainly by Calcar. It seems to us unlikely that any high proportion of the figures of the *Fabrica* and *Epitome* are by this artist. Who then made them ? As a working hypothesis, that at least covers the phenomena, we suggest that he was a young man, unknown to fame, who died before the book was printed.

But while the *Fabrica* bears on every page the stamp of the pioneer, it also carries innumerable marks of traditional scholarship. How can these diverse elements be reconciled ? If this question could be answered, we could at the same time resolve a main crux in the history of science and, indeed, in the history of the human spirit. But before even a beginning can be made in answering, it is necessary to examine an earlier work of Vesalius.

In the original form this work of Vesalius is devoid of title. It has become knonw sa the *Tabulae anatomicae sex* of 1538. In what follows we shall refer to it simply as the *Tabulae,* just as we shall speak of the *Fabrica,* the *Epitome* and the modern collection of Vesalian figures, the *Icones.* We have translated and annotated the *Tabulae,* so far as possible, and our task is now to supply such preliminary and commentatory matter as may help to render more intelligible both the *Tabulae* and its place in history. Success in this would mean throwing light on the rise of modern observational science.

The *Tabulae* is confessedly no more than a preliminary effort. Vesalius had a larger enterprise in mind even when its sheets were passing through the press. He ends his dedication with the words : " Should I find it acceptable to the learned I hope to add something more considerable thereto." But the same preface introduces an element of the Vesalian problem.

> "Not long since " he writes, " when appointed at Padua for the course of the Surgical part of Medicine, in discussing the treatment of Inflammation, I had . . . made a drawing of the veins . . . This . . . so pleased the professors and students of medicine that they pressed me for a similar delineation of the arteries and nerves. Since the conduct of dissections was part of my duty, and knowing this kind of drawing to be very useful to those attending the demonstrations, I had to accede to this request. Nevertheless I am convinced that it is very hard—nay, futile and impossible—to obtain real anatomical . . . knowledge from mere figures . . . though no one will deny them to be capital aids to memory."

That he should deem it necessary to defend the graphic method is a matter that, at the outset, needs some consideration.

§ 2. *Advent of the Graphic Method.*

In the Middle Ages anatomical figures were rare. The combination of representational incapacity with a degraded anatomical tradition made such few as existed well nigh useless. By the sixteenth century there had come, in anatomy as in other disciplines, a new reverence for the written word. For anatomy the authoritative sources were now the newly recovered texts of Galen. These were exactly edited, competently translated, and industriously studied, but they contained no figures. Despite the greatly improved standard of draughtsmanship and reproduction in the early sixteenth century, it would have been a formidable task to represent Galen's findings graphically. The *Tabulae* was the first real attempt to do so.

In the thirties, forties and fifties of the sixteenth century, however, an absurd controversy was being waged as to the usefulness of anatomical figures and physiological diagrams. The great contemporaries of Vesalius—Sylvius (1478–1555), Fernel (1485–1557), Guenther (1487–1574), Massa (*c.* 1500–69), Colombo (1516–59), Fallopio (1523–62)—did not employ them. To do so would be a break with classical tradition—or supposed tradition—in an age hypnotised by antiquity.* Figures suggested the amateur, the quack, the vulgarisateur. Sylvius, Fernel and others actively opposed their use. Among predecessors of Vesalius only

---

* Aristotle had used figures to illustrate the lectures for which his biological works were effectively the notes. This fact was unknown to or ignored by the participants in the controversy.

Berengario da Carpi from about 1520, Charles Estienne from 1532 onward, and Eustachio had the hardihood to employ graphic anatomical methods in their works.* Most of the figures of Eustachio (died 1574) remained unpublished till long after his death. All this must be viewed within the humanistic scene of an age which attached overwhelming importance to a rigid literary form.

But the *Tabulae*, after its gaseous and slightly fetid preface, almost entirely abandons literary form and passes to little more than a descriptive list of the lettering on the figures. It is the earliest attempt at a graphic exposition, to trained medical men, of both physiological and anatomical detail. Its first three plates contain the first pictorial exposition of the Galenic physiological system. Their method is a manifest break with tradition. The last three have a different character.

Vesalius indicates in his preface that he had prepared also a plate of the nerves. There is none in either of the existing copies of the *Tabulae*, but nevertheless it is not entirely lost. We know its general appearance from a pirated copy, or rather from a copy of a pirated copy.† It resembled a figure that afterwards appeared in the *Fabrica* ‡ (Figs. 3 and 4).

The last three tabulae have a different purpose from the first three. They are not organically connected with them ; they adopt a different method and are by a different hand. Vesalius writes in the introductory dedication :

> " Since many have vainly tried to copy what I have done, I have sent these drawings themselves [that is the first three tabulae] to the press. To them we have annexed ... three representations of my skeleton, which I had set up to the gratification of the students, rendered from the three standard aspects by the distinguished contemporary artist, John Stephen [van Calcar]."

Though these last three tabulae have been drawn on a principle different from and much older than that used for the first three, yet the six together establish a new form for fugitive anatomical sheets (see p. xxxiii). Very few, if any, of such sheets of a date earlier than 1535 have survived. Yet it is certain that these tabulae of Vesalius were found to be more useful than any of their predecessors, since plates imitating them were immediately and frequently printed at several centres.

Cushing gave much attention to such plagiarisms of the *Tabulae*. Neither he nor anyone else has shown that any surviving printed sheets of a date anterior to the *Tabulae* were copied from drawings made by Vesalius. Therefore we need not suppose that Vesalius, in the passage quoted above, referred to piratical *prints*, as has been generally assumed. It is much more likely that he is writing of hand-drawn copies privately exhibited anterior to March 1538,|| the date at which these words were written. In any event he is complaining only that the pirates have

---

* Estienne's work was begun in 1532 or earlier, substantially completed in 1539, but not published till 1545.

† Cushing, *Bio-Bibliography*, pp. 14 and 20 ; M. Holl and K. Sudhoff, *Des Andreas Vesalius' sechs anatomische Tafeln*, Leipzig, 1920.

‡ *Fabrica*, first edition, p. 219 ; second edition, p. 512. See Choulant-Frank, pp. 186–189 ; Cushing, pp. 20–21.

|| This is March 1538 ' old style '. We should call it March 1539.

copied his own tabulae, that is the first three. He makes no charge concerning the last three by Calcar.* This point will be evident if the passage quoted be read carefully.

Indeed it is only an infringement of the first three plates of which he could reasonably complain. The last three were the work and property of another. They are moreover purely representational. They are anatomical while his own are

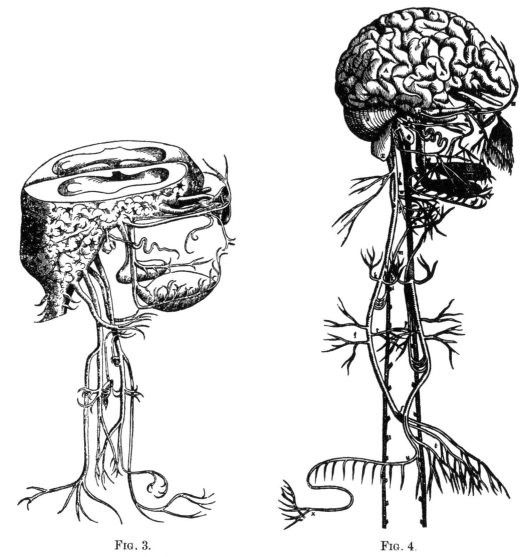

FIG. 3.   FIG. 4.

FIG. 3.—From a lost print by Aegidius Macrolios, published at Cologne in 1539 with the title 'The brain, fount and source of the Animal Spirit, communicating the action of the will (*sensum voluntarium*) to the whole body by the nerves that arise from it and from the dorsal medulla'. See Choulant-Frank, pp. 186–9, and Cushing, *Bio-bibliography s.v.* Macrolios. We have reversed the figure from its original for comparison with Fig. 4.

FIG. 4.—From *Fabrica*, p. 319. Brain and cranial nerves. Visceral branches of vagus not shown.

essentially physiological and the first of their kind. Representations of articulated skeletons had occupied some place in medical teaching even in the Middle Ages. Crude pictures of them appear in several early printed medical works,

---

* Johan Dryander (Eichmann) of Marburg pirated (1541) the drawings of the generative organs from the printed *Tabulae*. But there is evidence that some of Dryander's other illustrations were copied from drawings by Vesalius which were never published by him.

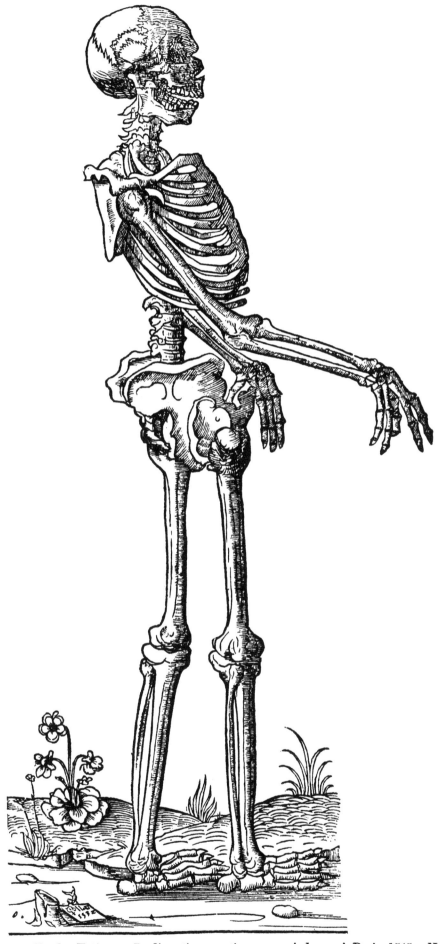

FIG. 5.—From Charles Estienne, *De dissectione partium corporis humani*, Paris, 1545. Near left heel the inscription Jollat stands for the name of the artist, Mercure Jollat. The 1532 shows that it was drawn six years before the publication of the *Tabulae*. It is one of a set of four representing views from front, back and both sides. We have removed the indicatory lines.

for example in Grüninger's edition of Brunschwig's *Buch der Cirurgia* (Strasbourg, 1497) and in Gersdorff's *Feldtbuch der Wundtartzney* (Strasbourg, 1517). But these were vernacular textbooks intended for unlettered surgeons. Such works were not in high medical repute in the sixteenth century.

As for the ' standard aspects ' of the skeleton to which Vesalius refers, they also have earlier counterparts. Thus Leonardo da Vinci drew views of a skeleton from back, front, and side, though he did not publish them. Giacopo Berengario da Carpi's *Isagoge breves* (Bologna, 1523) has figures of a skeleton from front and back. To Rosso de Rossi (1496–1541) is ascribed a plate of two skeletons and two corresponding muscle figures—one of each from back and front—which had been engraved by a pupil.* By some De Rossi is believed also to have designed the figures in the work of Charles Estienne. That anatomist displays careful studies of a skeleton from back, front and both sides in his *De dissectione partium corporis humani*, published at Paris in 1545. These figures, dated as drawn in 1532, anticipate those of Calcar by six years (Fig. 5). They were not printed till 1539, but their publication was yet further delayed by litigation.

Vesalius put together his first skeleton in 1536 at Louvain. That which is figured and discussed in the *Tabulae* however he articulated in January 1537 for a public dissection at Padua. It was of a youth of about eighteen. His age is shown both by the state of the epiphyses and sutures and also by the manuscript notes of a pupil of Vesalius.† It is not the same skeleton as that from which the three great skeletal plates of the *Fabrica* were drawn (Fig. 6).

It is clear that Vesalius intended that these six tabulae should provide illustrations for a general introduction to the study of anatomy and physiology, that is to the working (*fabrica*) of the human body. Their great size—in the original the skeletons are 17½ inches high—would make them suitable for such a purpose but inconvenient for the reading desk. They were doubtless designed to be suspended on a wall. There they could be used for demonstration during lectures or studied while reading such a text-book as the *Institutiones anatomicae* of Guenther, which Vesalius edited in the same year. They are primarily a student's aid for mastering the *Institutiones* of Medicine, that is the body of positive knowledge that should form the basis of the medical art.

In this capacity, as illustrating the *Institutiones*, the *Tabulae* fulfilled a new function. Its natural successor, the *Fabrica*, was to become immediately the oriflamme of a medical revolution and to provide a new way of looking at the body. Anatomy had been studied in some sort for centuries. A certain minor importance had long been attached to it. A hurried four-day demonstration on a cadaver had been an occasional or even a regular ceremony in some universities since the fourteenth century. Yet anatomy as a sustained study of structure was not an operative idea in medieval or earlier sixteenth-century medicine.

The thoughts of the physician until that time were expressed in terms of ' elements ' and ' humours ', ' qualities ' and ' complexions '. A vast tradition had developed concerning these and such like concepts. They were hypothetical entities, the separate existence of which no one could verify and, therefore, no one would deny. But, in the enduring human thirst for certainty, these intangibles,

---

* Choulant-Frank, *l.c.* p. 114.

† Vitus Tritonius Athesinus, on whom see M. Roth, *Andreas Vesalius Bruxellensis*, Basel, 1892.

imponderables and insensibles had become linked with the most positive of all the ancient disciplines—Astrology.

Ascendencies and descendencies of planets, the circling motions and patterns of the stars, eclipses of the greater lights and occultations of the lesser, all these could at least be exactly observed and recorded and as exactly predicted. The temptation to link the certainty of the heavens, the grand envelope of the Macrocosm, with the uncertainty of our earthly tabernacle, the poor abode of the Microcosm, was too strong to resist. By the sixteenth century medicine had become highly astrological. But the abrupt rise of a factual Anatomy, as embodied in the *Fabrica*, gave in a trice a new and far more positive basis to medical thought. In the course of a very few years medicine was to become anatomical. Observational science was thus born with the shortest of birth pains. The *Tabulae* of 1538 marks almost the beginning of the new birth.

Thus the graphic method in anatomy rendered high service in ousting judicial astrology from her ancient realm of spurious certainty. It is with the *Tabulae* that there opens the anatomic era of medicine, the age of graphic record of the arcana of nature, the systematically observational period of science. If we can display the sources of the *Tabulae* we may get a glimpse of one of the main roots of modern science.

§ 3. Art of the 'Tabulae'.

The *Tabulae*, as issued at Venice in 1538, took the form of six sheets, printed with the unusually large opening of $19 \times 13\frac{1}{2}$ inches. Our reproduction is but little more than half this diameter and therefore about a quarter of the size of the original. The text is arranged above and on either side of the figures. These figures, like those of the *Fabrica*, are true woodcuts, not engravings. They are printed, as are the letters of modern books, from ink that has been applied to raised ridges. Thus the black lines represent the ridges, the remainder of the general surface of the wood-block having been cut back with a sharp instrument. Shading is represented by lines more or less close together and often crossing each other.

It is difficult at first to grasp that it is the white spaces that are cut away in the blocks for the lines seem continuous and uninterrupted even when criss-crossed. The production of such elaborate wood-blocks demands great manipulative dexterity and long training. The technique had been practised even before the invention of printing with movable types. 'Block-books' are known from as early as the fourteenth century.*

In the late fifteenth century many woodcuts of great beauty were being made. Remarkable examples emanated from workshops at Venice, the chief printing centre of the age.† Fifteenth-century woodcuts are mostly confined to outline, shading being usually avoided. The impression of solidity was conveyed by the use of perspective and by variations in the thickness of outlines.

At the beginning of the sixteenth century, skill in the preparation of wood-

---

* H. M. Hind, *History of Woodcuts*, 2 vols., London, 1935 ; D. P. Bliss, *History of Wood Engraving*, London, 1928.

† Friedrich Lippmann, *Der Italiänische Holzchnitt in XV Jahrhundert*, Berlin, 1885 ; Prince d'Essling, *Les livres à figures vénetiens de la fin du XVe siècle et du commencement du XVIe*, Florence, 1908.

blocks had increased.   Method of shading by lines became more fashionable.   The art rose to its full magnificence in Germany where superb woodcuts of immense size were printed.*   Very fine blocks were also prepared in Italy towards the middle of the sixteenth century but few were superior to those of the *Fabrica* and *Epitome* of 1543.   The blocks for these two works, though made in Venice, were taken to Basel for imprinting.   That city had become the rival of Venice as a centre for book production.

The figures of the *Tabulae* are the largest ever employed by Vesalius, exceeding in height those of the *Fabrica* and even those of the *Epitome*.   They are among the largest ever prepared at Venice for any printed book.   Many of the actual blocks of the *Fabrica* have survived and have been reproduced in the *Icones*. None that were used for the *Tabulae* have come down to us.

We have no information as to who cut the wood-blocks for the *Tabulae*.   The first three were certainly drawn by Vesalius and we may safely assume that he employed a cutter to work from his drawings.   For the last three drawings Calcar may or may not have done his own woodcutting but he is not known to have prepared any other woodcuts.   Whatever the estimate of the draughtsmanship of the *Tabulae*—and it is certainly far below that of the *Fabrica* and *Epitome*—the woodcutting itself is good.   It is in contrast with the poor typographical style in which the text of the *Tabulae* is set forth.

The first three tabulae, being mere ' physiological diagrams ', are without artistic merit.   Their nature leaves little room for its display. Yet they reveal Vesalius as what we should regard as a very competent biological draughtsman. The shading is correct and intelligently applied, the light being consistently represented as falling from the right and above.   The curves are well taken ; enough detail is shown to illustrate the points described ; purposeless lines are avoided.   These diagrams are a pioneer attempt at biological draughtmanship so far as the printed page is concerned.   Thirty years earlier Leonardo had prepared many of a like kind, but none were printed and none were either as comprehensive or as well considered as these of Vesalius.

The three skeletons of Calcar, on the other hand, give scope for artistic enterprise.   Anatomically they are superior to and more detailed than any of their predecessors (those of Leonardo alone excepted), but this is not very high praise.   As works of art they lack both dignity and movement.   They contain many elementary errors and inconsistencies.   Thus, though the best of their kind to their day, they cannot be said to do much credit to a pupil of Titian.   To us it does not seem that they are by the same hand that drew the three magnificent skeletons of the *Fabrica*.   The jaunty gait of the fifth and the stiff lecture-room stance of the fourth and sixth tabulae are in strong contrast to the noble poses of the *Fabrica* skeletons.   Nor does it need much anatomical analysis to see that many elements in the *Tabulae* skeletons are out of drawing.   These conclusions seem to rule out the artist of the *Tabulae* as the artist of the *Fabrica*.

When the *Tabulae* was published Calcar was 39—an age which cannot be called formative for a draughtsman.   It is difficult to believe that one who could work no better in line could develop at 44 into the great creative artist of the

---

* See the prodigious work of Geisberg, *Der Deutsche Einblatt in der ersten Hälfte des XVI Jahrhunderts.*

Fabrica. For this reason, among others, we do not believe that Calcar made most of the drawings for the later work. Similarities in certain minute details of herbage and stones in the two sets of skeletons may be due to the same block-maker.

It might be thought that it would be a relatively easy task to draw bones accurately. Once the technique of graphic representation was mastered, it might be supposed that the portrayal of such structures, with their fixed and well defined outlines, would pass at once to its final stage. History shows, however, that this was not the case. With the exception of the drawings of the skeleton in the work of Estienne, those of the *Tabulae* are almost the first skeletal representations in a printed book which can be submitted to scientific analysis (but see Fig. 7). Apart

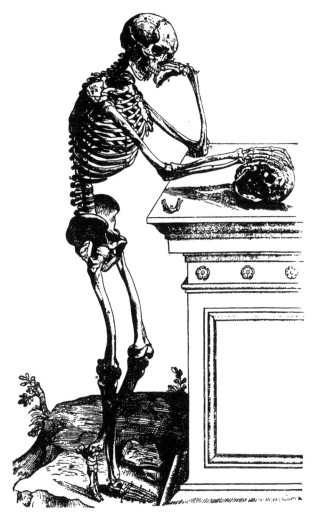

Fig. 6.—From the *Fabrica.* Contrast Tabulae IV, V and VI.

from the bony abnormalities due to disease, there are in these three skeletal drawings a number of features that must be ascribed to erroneous observation.

The worst-drawn bone is perhaps the scapula. Note especially the left scapula in Tabula IV. Both in the drawing and the text there is confusion between the acromion and the coracoid processes. The pelvic bones are throughout extremely badly rendered Note, for example, the exaggerated symphysis and the poor representation of the iliac crests in Tabula IV, the gross error in the pelvic tilt in Tabula V, and the enormous over-estimate of the width of sacrum in Tabula VI. The lumbar vertebrae are badly out of drawing, especially in the view from behind.

Errors have arisen from wrong articulation of bones.   Note the clavicles, both of which are shown wrongly articulated in Tabula IV, that on the right being back to front.   The right hand in this tabula is also wrongly jointed to the carpus.   The double curve in the spinal column is entirely missed in Tabula V.   In that figure also the very badly drawn scapulae are presented as alate.   The sternum is greatly misrepresented and is in seven segments.   Pelvis and coccyx are at the wrong angle to each other and to the axis of the spine.   The bones of the feet are wrongly articulated to the lower ends of the tibia and fibula.   The feet are completely devoid of arch.   Many other faults and errors might be indicated.

The skeleton is of a rickety young male.   All the long bones have irregular wavy outlines, the epiphyses are enlarged and the skull too large and globular.

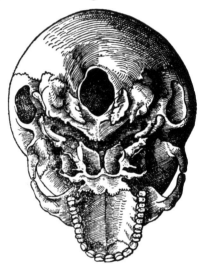

FIG. 7.—The best printed figure of a skull, and perhaps of any bone, before the *Fabrica*.   By an unknown artist from Balamio's Latin translation of Galen's *De ossibus* in Cratander's *Opera Galeni*, Basel, 1538.   Except for this work the edition is of the Greek text.   The *De ossibus* is reprinted from an edition of 1535, which we have been unable to consult.

These features are perhaps best seen in Tabula IV.   The age can be gauged from the epiphyseal junctions of the heads of the humeri, of the styloid processes of the radii, of the olecranon processes and of both ends of the femora, tibiae and fibulae. The age of 18, which we know from other sources, fits these points.

These three tabulae are below the standard that we should expect from a pupil of Titian engaged on a task of exact representation.   In our own day they would not increase the reputation of an artist.   But we know so little of Calcar, and most artists of the day knew so little of any anatomy beyond that of the surface muscles, that it is difficult to form any judgment as to the effect the work may have had on his contemporaries. Vasari notwithstanding, with these plates Calcar effectively disappears from the anatomic scene, nor does he, in doing so, shroud himself in a cloud of glory.*

* For Vasari's account of Calcar see Charles Singer, *Bulletin of the History of Medicine*, vol. xvii, pp. 429–431, Baltimore, 1945.

II. Vesalius and the Schools of Louvain and Paris.

§ 1. *Early Years of Vesalius.*

Remarkably little is known of the personality of Vesalius. There must be very few whose works created such a stir in their own time of whom contemporaries have less to say. For some periods in his life we have hardly any records. It is evident that he was not liked by many. If he was loved by any, their names have not reached us. An admirable account of nearly all the little known of him was given by Roth of Basel. To this Vesalian scholars must always turn. It is indispensable for its documentation but it fails to raise a living picture.*

Vesalius was born at Brussels in 1514. Learning was a tradition in his family. His great-great-grandfather Peter (*fl. c.* 1390) collected a medical library and edited a section of the Latin *Canon* of Avicenna. His great-grandfather Johannes (*c.* 1390–*c.* 1472) taught medicine at Louvain from 1429, was active as a mathematician, and wrote on calendar-reform. His grandfather Eberhard (*fl. c.* 1470) composed commentaries on the *Ad Almansorem* of Rhazes and on the *Aphorisms* of Hippocrates, together with mathematical works. His father, also Andreas, was apothecary to Charles V (1500–58, Emperor 1520–56), whom he followed on many journeys.

The life-course of all these men was centred in the Low Countries. In 1528 Andreas Vesalius began to study at Louvain. The university consisted then of a group of colleges or ' Pedagogia ' combined much as they are at Oxford and Cambridge.† These colleges at Louvain differed considerably from each other according to the degree to which they embraced or rejected humanism. Vesalius entered the conservative Castle College (*Pedagogium Castre*) ‡ where he received a grounding in scholastic Latin. Of Greek and of the newer classical approach he would have learned little, for a prejudice against the new humanism had developed there. He had, naturally, access to works on medieval science, and his boyish curiosity was specially aroused by reading the works on generation by Michael the Scot and Albertus Magnus. Faint traces of their views survive in both *Fabrica* and *Tabulae.*§

In after years—so it is said—he recalled how in those days he had a passion for dissection, and used to anatomise rats, moles, dormice, dogs and cats. In the early sixteenth century there were neither books on the subject nor encouragement for work of that kind. Such study implied strong individual taste.||

* M. Roth, *Andreas Vesalius Bruxellensis*, Berlin 1892. New light has been shed on Vesalius between 1549 and 1564 by Max Fisch, *Bulletin of the Medical Library Association*, XXXIII, p. 231 ff., 1945.

† Boys were not admitted to the Pedagogia till they were over 13. Vesalius attained this age in December 1527. On Louvain see H. de Vocht, *Alma Mater Lovaniensis*, Louvain, 1935.

‡ The name of this, as of other Pedagogia at Louvain, was drawn from an emblem or crest carried by the members.

§ *Fabrica*, 1543, p. 531.

|| The account of his early anatomical efforts reaches us very indirectly in Theodore Zwinger's *Theatrum vitae humanae*, Basel, 1585. Zwinger (1533–88) of Basel was nephew of Oporinus, printer of the *Fabrica*. After his father's death his mother married the scholarly Conrad Wolffhart (Lycosthenes, 1518–61), professor of dialectic at Basel. Zwinger, though only ten, must have come across Vesalius when he visited the place in 1542–3. Wolffhart spent the last 15 years of his life on the anecdotal collection *Theatrum vitae humanae*. His step-son, Zwinger, completed it.

About **1531, when** Vesalius was seventeen, he transferred to Busleiden's *Collegium trilingue* at Louvain. It was there that his literary attitude was determined. The educational ideal for which this college had been founded was that of the *trilinguis homo*, the man of three languages. The phrase has a meaning peculiar to the time and place. In the concept of Erasmus (1467–1536), embodiment of the humanistic spirit, a *trilinguis homo* was one who had obtained full entry into Latin, Greek and Hebrew. To us these seem very inaccessible fastnesses but the humanists of the sixteenth century lived in an atmosphere almost entirely made up of language. This was their great weakness and was satirised by Erasmus himself in his *Ciceronianus* (1528), directed particularly at the pedants of Padua. The trilingual college at Louvain had been established in 1517 by the enthusiastic humanist Busleiden (1470–1517), friend of Erasmus and of Sir Thomas More. He desired to stamp on its scholars what he regarded as the three emblems of civilisation, the three classical tongues. Did he succeed with Vesalius ?

The spirit of the place impressed itself on Vesalius. His restless temper naturally prevented him from becoming a great scholar, but at Busleiden's college he came to understand humanistic scholarship in its trilingual aspect. It is improbable that he ever gained facility in Greek and he certainly knew no Hebrew, but nevertheless he became in spirit a trilingual humanist.

Moreover he united this humanist aspiration to the pure spirit of scientific research. In that he stands almost alone. No *trilinguis homo* surpassed him in scientific achievement. No early man of science surpassed him in philological appreciation. Whatever his competence as scholar, it is yet true that he was interested alike in Hebrew traditional learning, in the Arabian scientific legacy, and in the recovery and improvement of Greek texts. With all these interests we shall be here concerned.

In the course of the sixteenth century all colleges at Louvain became centres of scholastic reaction but, during the formative years of the young Vesalius, Busleiden's college, though it had lost its original eminence, was still among the homes of trilingual learning. As to the relation of Vesalius to the general life of the university we know nothing, except that he was dissatisfied with its training. At this time probably he met Narciso Verdun, to whom he afterwards dedicated the *Tabulae*. Narciso was in Flanders at the court of Charles V in 1532. He was presumably in contact with the Emperor's apothecary, the father of Vesalius. (See note 1 to Tabula I.)

In 1533, on the advice of a medical friend of his **father,*** Vesalius moved to Paris. The full classical discipline in medicine had just become available there. Galen had begun his long Lutetian reign. The Arabian writings were falling into neglect. Leading Parisian teachers of Galenic Medicine were Sylvius (p. xvii) and Guenther (p. xvi). To them flocked some of the ablest students of Europe. As with all eminent teachers of that age, we read that their audiences ran into hundreds. No contemporary drawing of a medical lecture at any university, except the title-page of the *Fabrica*, shows such numbers, and Vesalius did not include meiosis among his more habitual figures of speech (Figs. 1, 2 and 8).

To understand the climate that the young man now entered, it is necessary to know something of the traditions and history of the Parisian anatomical school.

* Nicolas Florenas, to whom he dedicated his first work, *Paraphrasis in nonum librum Rhazae . . . ad Almansorem*, Louvain, 1537.

§ 2. *Anatomy at Paris.*

The University of Paris was extremely conservative and the Faculty of Medicine at least as resistant to change as the other departments. The early history of anatomy at Paris has not been exactly traced. The medical school building in the Rue de la Boucherie was completed in 1477. No department in it was designated for dissections. These rare ceremonies were commonly held in a cellar of the hospital Hôtel-Dieu. From 1483 a formal demand for a knowledge of anatomy was made on candidates for the baccalaureat but no dissection was held till 1493. Dissections became more frequent after an appeal of the Medical Faculty to the Parlement in 1526, but they remained comparatively rare until after Vesalius left in 1536.

Parisian anatomy was backward in this respect even by medieval standards. Paris had little contact with the great Italian universities. There was no authoritative Parisian anatomical text. Neither Mundinus nor any of the Italian anatomical authors was studied. Anatomical reading was confined to matter buried in surgical works, but surgical instruction was relegated to the Surgeons' Hall of St. Come and St. Damien. Thus when Galen's anatomical texts became generally accessible at Paris they had no competitors. The advent of the renaissance element was later in coming to Paris than to Padua, but its victory was the more dramatic. Galenism was very rapidly established and soon became itself part of the traditional conservative order. It was not displaced till after the seventeenth century and then but slowly. Moliere (1622–73) was very familiar with it (*Le malade imaginaire*, 1673).

The new classical scholarship was introduced at Paris largely by Guillaume Budé (Budeus, 1467–1540). Its opening event was the publication of his famous *De asse* (1514) in the very year of birth of Vesalius. The tradition entered the medical sphere through Jean Ruel (Ruellius, 1479–1537). He was very learned in Greek medical texts and occupied an influential position in the medical faculty but did not interest himself in Galen. There was a first Galenic ripple at Paris in 1514 when the scholar-printer Henri Estienne (died 1520) issued a small collection of texts of Galen translated by the venerable Nicolo Leoniceno of Ferrara (1428–1524), then in his eighty-sixth year. The tide rapidly gathered strength.

Galenic texts were soon pouring from the presses of Paris. The rate had reached about two a month by 1528. In that year Simon de Colines, who had married Henri Estienne's widow, issued Latin translations of four texts of Galen which provided a basis for the new anatomical teaching. They were (a) *De usu partium corporis humani* in the 14th century translation by Nicolas of Reggio; (b) *De motu musculorum* in the translation of Leoniceno * ; (c) *De facultatibus naturalibus* rendered by the great English scholar Thomas Linacre (1460–1524),† which gave standard authority to the current Galenic system of physiology (see p. xxxviii); and (d) *Introductio seu Medicus*, a convenient pseudo-Galenic pamphlet interpreted by Guenther. All four are little pocket volumes designed for students. Three years later (1531) Guenther translated in a handsome tome the first nine books of Galen's *De anatomicis administrationibus* (Fig. 8).

An academic event that fitted well the Louvain tradition of Vesalius took

---

* An edition of this appeared in London, where Leoniceno was known personally, in 1522.

† This was a reprint and had already appeared in London in 1523.

place at **Paris in 1529.** In that year Francis I (reigned 1515–47) founded the Collège de France for Greek and Hebrew studies. In doing this he was influenced by his sister **Margarite d'Angoulême,** by Budé and by the De Bellays. The college was to continue the work that Lascaris (p. xviii) had already started in the Royal Library at Fontainebleau. It was on a similar plan to that of Busleiden's college at Louvain (p. xiv). Vesalius was thus no stranger to these ideas when he arrived in 1533, a youth of nineteen.

Certain medical teachers at Paris also fitted this scheme. Among them was Johannes Guenther (1487–74) of Andernach, on the left bank of the Rhine. He was educated in the Low Countries and had taught Greek at Louvain before Vesalius entered there.* He came to Paris in 1527, just when the Galenic tide was beginning to set strongly, and he became a popular teacher. As translator he was very industrious but over-hasty. His anatomical demonstrations, which Vesalius attended, were said by his pupil to be contemptible. They lasted only three days

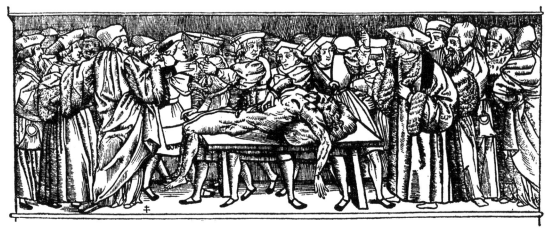

Fig. 8.—Panel showing dissecting scene from the title-page of Galen's *De anatomicis administrationibus,* translated into Latin by Johannes Guenther of Andernach and printed at Paris in 1531 by Simon de Colines. A crowded room contains about twenty-five figures in academic dress. On a table in the foreground lies a body of which the abdomen has been opened. A discussion is in progress. A young enthusiast is excitedly demonstrating, with one hand in the abdomen and the other raised in gesture. An older lecturer stands at the head of the dissecting bench. He is making a point. Another youth hands some of the viscera to a yet older academic. There are no books in the scene. Contrast with Figs. 1 and 2 (see p. ii).

and, despite his Galenism, were medieval in character. They were indeed below the standard of Mundinus. for Guenther's dissections were the work of ' demonstrators '.† So far as the abdomen was concerned they consisted of little beyond the conventional separation of the muscular coverings and a general display of the viscera. The stress on the muscles of the abdominal wall was a legacy of medieval anatomy.

But Guenther's Galenic teaching naturally appealed strongly to Vesalius. Guenther summarised it in his *Institutiones anatomices secundum Galeni sententiam* (Simon de Colines, 1536), a brief survey for candidates in medicine. It influenced anatomical nomenclature deeply. Many of the terms which were thus popularised

* It is sometimes stated that Guenther taught Vesalius at Louvain. The dates of the two men render this impossible.

† *Fabrica,* p. 3 recto, line 38 ff.

had already appeared in the works of Pollux (1502, see p. lxviii), Benedictus (1514, see p. xxix), Valla (1527, see p. xxix), or in Guenther's own version of the pseudo-Galenic *Introductio seu Medicus* (1528, see p. xv).

The other great medical teacher during the Parisian sojourn of Vesalius was Jacobus Sylvius (Jacques du Bois, 1478–1555). He began medical study so late that, though older than Guenther, he was junior to him on the Faculty. His earlier years were devoted to linguistic and humanistic studies. He was 51 when he took his medical degree in 1529, and he did not begin to teach in the medical school till 1531, two years before the arrival of Vesalius. Despite his late start he became an extremely successful and popular instructor.

We must not look at Sylvius entirely through the yellowish spectacles of Vesalius. The talents of the two men were so different that mutual understanding was hardly possible. Vesalius was a discoverer. His expansive, ardent genius did not make for the orderly arrangement of the great territories of knowledge that he won. An explorer cannot delay to perfect his maps. Moreover Vesalius was devoid of literary capacity or, rather, of literary charm. Least of all had he the gift of succinct and well arranged statement. Sylvius, on the other hand, was essentially a systematiser with a special gift for the ordered and concise marshalling of knowledge. His *In Hippocrates et Galeni physiologiae partem anatomicam Isagoge,** representing his teaching from about 1531, is admirably clear, brief, and exact. It is obviously modern and neither ancient nor medieval, and is just the kind of teaching that Vesalius needed. Nor is it merely Galen and Hippocrates retold but a good piece of systematisation, tacitly omitting much from the ancient texts that is contradictory or erroneous. It introduces the modern method of numbering branches, organs and relations. Some parts might almost replace certain sections of our students' textbooks. Regarded simply as a feat of verbal exposition, it is superior to anything produced by Vesalius. In this *Isagoge* Sylvius provided the student with a very skilfully constructed scheme of the muscles which must be regarded as the foundation of modern muscle nomenclature. It was based, as was Galen's, on their attachments. It also popularised a number of Latin or Latinized names for vessels. Many have persisted to our time (see p. lxxi). In view of subsequent controversies it is interesting that this book states that the ancients dissected monkeys rather than men and that there are differences in the structures of the two types.

Nor must Vesalius be taken too literally when he suggests that he was the first 'to put his own hand to the business' in the matter of anatomy. Guenther and Sylvius are doubtless here unfortunate contrasts to Vesalius himself, but before the time of Vesalius, as well as in his time, there were learned anatomists at Paris who were actively dissecting. On the very title-page of Guenther's version of the Galen's *De anatomicis administrationibus* of 1531 is a vivid anatomical scene. Parisian doctors in academic dress throng about the body. One has very much 'his hand to the business', for it is thrust in the abdominal incision while a colleague passes some of the organs to another (Fig. 8). Nor do the activities of such men lack a literary memorial. There is the work of Charles Estienne.

---

* First printed Paris, 1542. There are many editions, and there is a French translation of 1555.

§ 3. *Charles Estienne.*

A place in a discussion of the *Tabulae* must be given to the anatomical treatise of Estienne (1504–64). It is the most notable of its kind from the Parisian school. Estienne's career and achievement are contrasts to those of Vesalius. As son of the learned printer Henri Estienne (1460–1520) he had a better training in the classics than Vesalius. Budé (p. xv) was a close friend of his family.

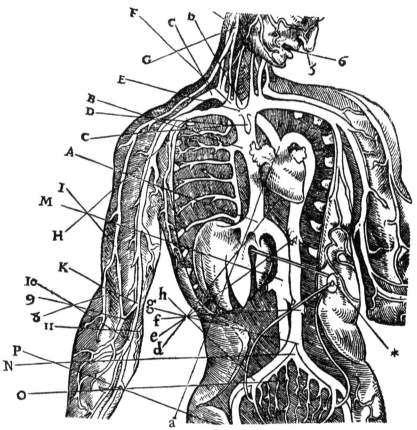

FIG. 9.—Vascular system according to Estienne (before 1539), enlarged from original. *Truncus brachiocephalicus* and *vena cava superior* are of ungulate type. The two venae cavae are directly continuous as one trunk of which the right atrium is a main branch. The liver, somewhat like that of a cat in form and position, is shown as the source of the caval system. The right renal vessels are higher than the left. The azygos, as with Vesalius, enters the *vena cava* on its right instead of behind. Capital letters indicate 'divisions' of *vena cava*, small letters those of aorta, numerals show places for phlebotomy. C represents the beginning of the internal mammary arteries. G is the junction of a conspicuous branch of the cephalic vein with a large external jugular which forms the main jugular trunk as in ungulates. H is a large external thoracic vein as in ungulates and many other animals. At * conspicuous vessels converge on the umbilicus. These are lateral umbilical ligaments and are shown and described as functional arteries. The straight vessel that runs from umbilicus to liver is the vestige of the umbilical vein. Vesalius had possibly seen this figure by 1538. See note 60.

The learned Greek A. J. Lascaris (1445–1535) was his teacher. Simon de Colines was his step-father. He is said to have taken early to medical studies, though he did not receive an M.D. degree until 1542, six years after Vesalius, who was ten years his junior. Charles Estienne had, however, probably begun anatomical research in 1530, and he was certainly occupied on it in 1533, when Vesalius arrived in Paris (Fig. 5). Vesalius concentrated terrific energy on one great work, Estienne diffused his powers in many fields. He was a poet, wrote much on various subjects, including education, history, Latin and Greek grammar, road-systems, hunting and agriculture, practised as a printer and died in a debtor's prison. His name appears on the title-pages of no less than seventeen separate and very diverse works.

Estienne's *De dissectione partium corporis humani*, 1545 (French edition translated by himself 1546), was printed by his step-father. In a short preamble ' to his students of anatomy ' he says that ' this work was almost finished by 1539, and practically all printed as far as the middle of the third and last book, when we were forced by a lawsuit to abandon its completion '.* Certain of the earliest plates contain details of the surface-musculature that would be of interest to an artist. The suggestion that they are by Rosso de Rossi has been made. The book was actually in print within a few months of the *Tabulae*, but there is no trace of direct influence of Estienne on Vesalius or of Vesalius on Estienne. Statements to the contrary are, we believe, based on misunderstanding. It is very unlikely that the two men were intimate in Paris. Vesalius was then an eager student in his late teens, Estienne an eccentric dilettante of thirty.

In his preface Estienne states that many of the illustrations and dissections were by the surgeon Estienne de la Rivière, with whom he worked for several years. Several are signed by Mercure Jollat (Fig. 5), but only one bears the initials S.R. (Stephanus Riverius = Estienne de la Rivière). The two Estiennes quarrelled over the book. This is not hard to understand of either.

Estienne's book is the ugliest anatomical work that we know. The anatomised areas are mostly represented by blocks clumsily inset in figures skilfully and dramatically posed, but wholly devoid of grace or beauty. The insets, too cramped for convenient study, often increase the hideousness of the contorted shapes and help to make the book an unpleasant companion. We shall deal later with certain of its resemblances to the *Tabulae*. (See Figs. 5, 9, 11, 12.)

§4. *Vesalius at Paris*.

The Parisian teachers regarded medicine integrally ; anatomy had not reached the dignity of a separate discipline ; dissection was backward ; surgery occupied an inferior status. The ruling medical humanists were grappling with the task of translating and commenting on the great corpus of Galen's writings. They had now reached the books of anatomic content. The *De usu partium*, the *De facultatibus naturalibus* and the *De anatomicis administrationibus* had all become accessible shortly before the arrival of Vesalius. The last work had just been translated by Guenther (1531, p. xv). It is devoted to practical details of dissection and vivisection. Its vivid descriptions aroused much less specialised minds than that of Vesalius. They drew Fernel (1497–1558) from his astronomical speculations † and Charles Estienne (1502–64) from his flower-garden ‡, the former to become the coryphaeus of medicine, the latter to anticipate Vesalius as an active anatomist. The calm, philosophic, eloquent Fernel became the best exponent of physiological views of the century. § The unstable temperamental Estienne produced the only anatomic work of the time comparable to those of Vesalius and Eustachius.

---

* Of a total of 406 pages in the French edition, Book III contains only 64 and these are manifestly inferior in character to the earlier part. They contain a barren description of muscles detached from the bones and a poor account of the spinal cord which, however, records its central canal.

   † Jean Fernel, *Cosmotheoria*, Paris (Simon de Colines), 1528.

   ‡ Charles Estienne, *De re hortensi*, Paris (Robert Estienne), 1535.

   Jean Fernel, *De naturali parte medicinae*, Paris (Simon de Colines), 1542.

There is no need to take too seriously the later account by Vesalius of his revolt against the Paris teaching. He was still receptive. He still trusted in the written word as firmly as any Italian pupil of Mundinus. True he had better texts. The assertion of the claims of the Greeks was all around him. One who posed as a trilingual man would be the last to doubt that Galen stood above Avicenna. No humanist would doubt that. But humanists in interpreting nature stood quite as firmly by the books of the Greeks as did the scholastics by those of the Arabs.

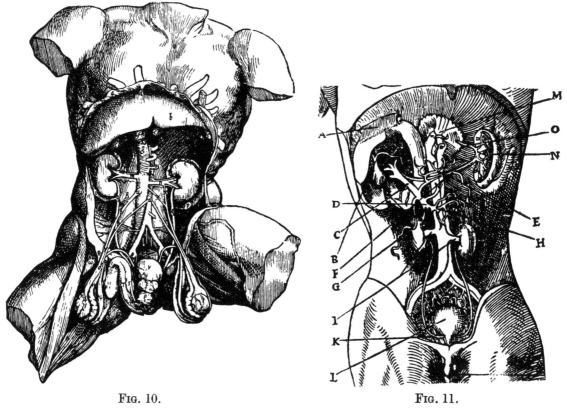

FIG. 10.   FIG. 11.

FIG. 10.—From *Fabrica*, p. 372. Right kidney and renal vessels higher than left as in animals, though the right side of the liver is shown as larger than left as in man. Origins of spermatic arteries and veins well seen.

FIG. 11.—Posterior wall of abdominal cavity from Estienne (before 1539). A is the suspensory ligament of liver; B, cystic duct; C, portal vein, and D, F, G, H its branches; E, branch of portal vein to stomach; I, *venae emulgentes*; K, ureter; L, bladder; M, 'end of œsophagus cut off near diaphragm'; N, spleen; O, 'vessel from the spleen carrying melancholy humour to the upper orifice of the stomach'.

Vesalius when he first 'set his own hand to the business' (1534–5) did so to verify Galen. How could he possibly do otherwise ?

In after years Vesalius said that he was disappointed with Paris. Guenther and Sylvius, he said, were almost indifferent to dissection and were enthusiasts only for what Galen had written. But in 1542, when Vesalius wrote, ' I could never have reached my end in Paris had I not put my own hand to the business ',* he forgot several important factors. Firstly he forgot that had these things been otherwise, he would not, as student, have had that freedom of dissection. Secondly he forgot how much he owed to the clarifiying effect of systematic expositions of Guenther and Sylvius. And he forgot something yet more important. He forgot that for five years after he first set his own hand to the business he was almost as

* *Fabrica*, Preface, p. 3, line 26.

tied to the written word as Guenther and Sylvius. The *Tabulae* shows Vesalius as a typical renaissance trilinguis homo, though with the added talents of an intensely visual mind and an enthusiasm for dissection. He is still only 23 ; he is still tied to Galen ; he is not yet a discoverer. But the' *Fabrica* reveals him as the very type of the modern man of science, with a rovingly enquiring mind and the habit of presenting his own findings in pictorial form. The metamorphosis from the one type to the other took place within the intervening four years. 1538–1542 is the period of gestation for the idea that science is research.

Sylvius was wont to speak of a ' book of Nature '*—a cant phrase. Nature was for him too literally a book—that of Galen—which his pupils should read and learn. For Vesalius, even in his Parisian stage, things must have looked a little, but only a little, different. He treated the written word as a traveller might a guide-book. He explored, with more enthusiasm than others, the territory described. Nevertheless he used the same book and seldom made an excursion not laid down in it.

It seems that at Paris Vesalius dissected many animals and examined such human material as he could. The most readily obtained human specimens were, of course, bones. His persistently investigatory spirit accumulated a knowledge of these far beyond anything in medieval works, but bones had naturally always been better described than other parts. Vesalius claimed that he learned to recognise blindfold all the bones by exploring them with his hands.† Too much should not be made of this not very difficult accomplishment, for gross osteological errors survive in the *Tabulae*.

His talents soon won recognition. He tells that at the second anatomy which he attended (1535) he was asked by teacher and students to assist ; that he proved his superior skill ; and that at the third anatomy (1536) he did the work almost unaided. Vesalius was never one to hide his light under a bushel. He demonstrated the contents of the abdominal cavity and the musculature of the arm far better, he says, than had ever been done before.‡ Some years later (1539) he traced to this occasion his first doubt as to the accuracy of a detail in the text of Galen,§ but the doubt could only have been a very little one and it is not very apparent in the *Tabulae*. Moreover the *Fabrica*, and especially Book III, which is on the vascular system, is full of Galenic findings in animal bodies.

In 1536, just before Vesalius left Paris, Guenther published his *Anatomical Institutions according to the opinion of Galen for students of medicine*. It gives an idea of the teaching which Vesalius had received and of the anatomical knowledge that he took away with him. He certainly assisted Guenther with this book. It was, moreover, the first that he was to edit after the *Tabulae*, which is, in some senses, a companion to it. It is a short work of purely Galenic content. It almost excludes not only the Arabs and Arabists but also Aristotle, as well as the medievals and moderns. It pays lip service to observation. We have it from Guenther's own mouth that it was Vesalius who did any dissection needed for it. ‖ Guenther indicates that Vesalius first demonstrated the origin of the spermatic vessels (but see p. lxi), though he omits to mention him in connection with the azygos vein.

* Jacobus Sylvius, *De ordine et ordinis ratione in legendis Hippocrates et Galeni Libris*. Paris, 1539.

† *Fabrica*, p. 159, line 24 ff.

‡ *Fabrica*, Preface, p. 3, line 32 ff.

§ *Epistola docens venam . . . secandam*, Basel, 1539, p. 29.

‖ In extreme old age, and after the death of his pupil, Guenther wrote : ' When I presided at a public dissection at Paris Andreas Vesalius did the work for me ', *De Medicina veteri et nova*, Basel, 1571, I, p. 159. Guenther was 84 when this book was published.

To the end of his Paris days, and beyond, Vesalius remained a Galenist. In the Introductory letter to the third edition of the *Institutiones* (Basel, 1539) Guenther expresses his thanks to Vesalius and Servetus for their anatomical demonstrations and admiration for their Galenic learning. We have no further record of the association of these two strangely contrasted young men.

§ 5. *Vesalius returns to Louvain.*

The medical school at Paris dispersed in July 1536 when Charles invaded Provence. Guenther and Vesalius left Paris : Vesalius for his native country ; Guenther, whose position in Paris was uneasy because of his Protestantism, for Wittenberg, its centre. Vesalius was 22. His reputation for anatomical skill must have reached his old university. A physician at Louvain asked him to make an

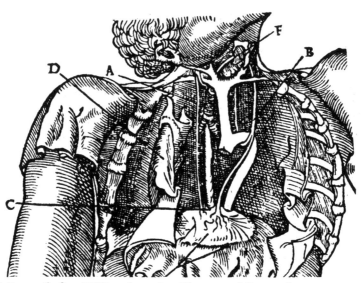

FIG. 12.—From Estienne (before 1539), enlarged. The opened thorax shows a *truncus brachiocephalicus* of ungulate type given off from the aortic arch. ' A, place of return of that part of the sixth nerve [our X or vagus], called recurrent. At this spot, which is under the clavicle, the nerve is turned round the branch of the artery of the right arm ; B, place of return of the corresponding structure on the left side but much lower than the other and in the neighbourhood of the division of the great artery, of which the larger branch turns downward ; C, oesophagus passing through the diaphragm to go to the stomach ; D, coat which surrounds the ribs within and known as " pleura ".'

autopsy on a case of suspected poisoning in a girl of eighteen. This was his first independant post-mortem examination and the first female body that he had opened. He found a *corpus luteum* in the ovary. It made a strong impression on his mind, for he described it in detail twenty years later.*

About the end of 1536, seeking bones in the place of execution at Louvain, he found hanging on the gallows a skeleton held together by the remaining ligaments. He brought it in almost intact, securing the few missing bones in subsequent expeditions. This was his first complete skeleton. The procedure was less suspect than might be thought. The burgomaster himself, hearing of his enthusiasm, removed any obstacles to his osteophilic activity.

The students of Louvain were eager enough for his anatomical instruction. At the beginning of 1537 he gave the first demonstration of human dissection seen

* *Fabrica*, second edition (1555), p. 658, line 20 ff.

there in eighteen years. He lectured on the body while he dissected. This was at once a reform and a return to Mundinus (1316). But he did not usurp the professorial function, for this anatomy was conducted in the medieval manner, as a ceremony with a senior acting as 'president' (compare Figs. 1 & 2). Fortunately that officer, far from holding himself pompously aloof, was friendly to the progressive young teacher.*

The career of Vesalius was throughout relatively free from authoritarian interference, clerical or other. It might be thought that one so impetuous and combative, so obviously sceptical, and working on a suspect subject, would encounter obstruction and persecution in Paris, Louvain, and Italy. The reason for his immunity is not however obscure. He was a researcher in the modern scientific and limited sense. He took little interest in great general ideas, none of which was he pledged to refute. He had but small philosophic power and there is no evidence that he had much philosophic interest. He never exhibits any but conventional sentiments on ' religious ' themes in his books. He disliked priests and monks and said so often enough, but such distaste was very far from uncommon and would not in itself bring him serious trouble. The medical profession of those days, as of these, was generally only too ready to avoid theological entanglements. And the one ruling philosophic conception of Vesalius, that Man was made in the image of God, suited peculiarly well the intellectual climate of the great movements in which his life was cast—humanism, renaissance art, Catholic philosophy, and the ideas of reformation and counter-reformation. But his companions he could and did choose at will from Catholics, Protestants and Jews. Padua was the best place in Europe to find the varied intellectual society that suited his temper.

Nevertheless, as we follow his career, we do at times catch a distant clerical growl. At his first anatomy at Louvain certain theologians took offence at his view of the seat of the soul. One confronted him with an antiquated picture from the well-known *Margarita philosophica*.† It is a work by a Carthusian prior of Friburg and is an encyclopaedia on typical medieval Aristotelian lines, illustrated by the coarse and hideous figures characteristic of the German press of the day. This ridiculous work was put up to support Aristotle's *De anima* against the supposed views of Vesalius. He had the sense not to enter into discussion, but the matter rankled with him for years.‡

Of greater importance for his anatomical career than these philosophic passages of arms was an acrimonious debate on venesection. The object of his attack, Jeremiah Driviere (1504–54), had taken a degree in philosophy at Louvain, and was now devoting himself to medicine. In 1532 and 1535 Drivière published two controversial works on blood-letting. § These maintained, with the Arabians, that blood should be let on the side opposite to the pain. This was contrary to what was held to be the Hippocratic view. In this barren dispute he took from his early days what he regarded as the Hippocratic side. He held that

---

* To this president, Johannes Armentarianus (Heems), Vesalius dedicated his revision of Guenther's *Institutiones*, Basel, 1538. He was perhaps a son of Nicolas Heems (c. 1470–1532) of Brussels, who long held the chair of law at Louvain.

† By George Reisch. The first illustrated edition is that of Friburg, 1502.

‡ *Fabrica*, First Edition, p. 623.

§ *De missione sanguinis in pleuritide*, Louvain, 1532, and *De temporibus morborum et opportunitate auxiliorum*, Louvain, 1535.

FIG. 13.—Frontispiece of *Fasciculo di medicina*, Venice, 1493. A university lecturer at Padua is speaking from his ' chair '. The range of his medical knowledge is shown by the volumes displayed. Above are the great classics : the Greeks, Aristotle, Hippocrates and Galen, and the ' Arabs ', Avicenna, Haly Abbas, Rhazes, Mesue and Averroes. On a revolving desk for reference is Pliny's *Historia naturalis*. Other books are seen in the cupboard. Below them is the *Conciliator* of Peter of Abano, the works of Isaac the Jew, and a volume of Avenzoar. In the foreground are sick people, each with a carrier for a urine bottle.

bleeding should be from the vessel nearest the lesion. This proximity, according to him, should be determined for the upper intercostal spaces by the peculiar distribution of their veins in relation to the azygos (see p. lvi and note 58). The discussion, in the light of our Harveian physiology, is quite empty. Nevertheless it gave rise to a whole library of works. *Odium theologicum* was among the weapons wielded by the windy logothetes in this gaseous warfare. On it Vesalius, unfortunately, spent much energy and time.

In February 1537 Vesalius printed at Louvain what was doubtless his graduation thesis, his *Paraphrasis in Nonum Librum Rhazae ad Regem Almansorum*. Of it no perfect copy has survived. He reprinted it at Basel a few weeks later. The book is unimportant and without originality. It is significant here only for containing the *suggestio falsi* that Vesalius could read and translate Arabic (p. lxxvii). On leaving Basel he went to Italy, where the *Tabulae* was printed in the following year.

§ 6. *Anatomical Equipment of Vesalius in 1537.*

The anatomical experience and reading of Vesalius on his arrival in Italy may now be summarised.

On the practical side he could draw on (*a*) experience of human bones and of a complete skeleton; (*b*) frequent dissections of several species of animal, notably of the dog, and public demonstrations of animal organs by Sylvius; (*c*) a small number of public anatomies, at several of which he had himself acted as demonstrator; (*d*) knowledge of human organs picked up at places of execution etc; (*e*) a few post-mortem examinations.

On the theoretical side he had read the current Latin versions of the Arabian writers Avicenna, Rhazes and probably Haly Abbas (Fig. 13). Unlike Sylvius and Guenther, he retained his interest in these. This is shown by his *Paraphrasis in Nonum ad Almansorem* (1537), by very many references in the *Tabulae* (1538) and by his treatment of Arabic and Hebrew nomenclature in the *Fabrica* (1543). It is unlikely that he had yet read any medieval or modern Italian anatomical work, nor is there evidence that he learned anything from the anatomical activities of Charles Estienne, who was dissecting from about 1531. His great asset was his mastery of Galen's anatomical and physiological writings. We must consider which of these were accessible to him.

He was, of course, intimate with the Galenic anatomical texts circulating in Paris (see pp. xv–xvii). By 1537 there had also appeared four considerable collections of Galen's writings, three in Latin and one in Greek. The Latin collections are the first Venetian (1490), that of Pavia (1516), and a repetition of the latter, issued by the house of Giunta at Venice in 1522. (This last is usually disregarded in numbering the Giuntas.) The Aldine or first Greek edition was printed at Venice in 1525. The Kratander or second Greek edition was in preparation in 1536 and was issued in Basel in 1538.

In the first Venetian edition of Galen (1490) there was no important anatomical text. That subject was represented in it only by the medieval *De juvamentis membrorum* abbreviated from the *De usu partium*, (p. xxvi). In the Pavia edition (1516) the *De usu partium* appears in the old translation of Nicolas of Reggio (p. xv). With both these texts Vesalius was very familiar. He had read them at Paris in separate issues, so that the great collected editions were of little help to him. The Aldine Greek edition (1525) has all the important

e

anatomical and physiological texts of Galen except the *De ossibus* (*v. infra*) which first appeared in a Latin translation at Paris in 1536, and again in Latin in the Greek collected edition of Kratander at Basel in 1538.

Vesalius would be unlikely to profit by any untranslated Greek text. There is no evidence that he ever attempted to do so. He lived for eight months in the same house as that ardent Grecian, John Caius. All that he tells of the scholarship of Vesalius is that certain notes on Greek texts were useless to Vesalius and to himself.* Had the two men similar philological capacity Caius would surely have had more to say of his companion. The Greek scholarship of Vesalius demands a special investigation.

In addition to the collected Galens there were separate works, many translated by Guenther. Vesalius was naturally familiar with these. Sylvius, on the other hand, despite his reputation as a scholar, hardly ventured independently into the field of translation from the Greek. He mostly confined himself to abstracting, arranging, annotating and systematising ancient works. It seems that he was much helped in such matters by another member of the medical faculty, Martin Gregoire of Tours, whose Greek learning was well considered.

The more important Galenic texts with which Vesalius was intimate may be arranged in the approximate order of their significance for his subsequent work. To each we give a reference to Kühn's collection (*Galeni Opera Omnia*, 20 vols., Leipzig, 1821–33) as the standard modern edition and to the 'second' (really third) Giunta (in 7 'Classes' or Vols., Venice, 1550) as the standard renaissance edition. This has an admirable index by Brassavola.

(a) *De anatomicis administrationibus*, first nine books (Kühn, II, p. 215; Giunta I, f. 63), Latin by Guenther, Paris, 1531. The work left the deepest impression on Vesalius. He revised Guenther's version in 1541. There are many references to this work of Galen in Guenther's *Institutiones*, 1536, in its revision by Vesalius of 1538, in the *Tabulae*, 1538, in the *Fabrica* of 1543, and in the *Epistola . . . radicis Chynae*, 1545.

(b) *De usu partium corporis humani* (Kühn, III p. i; Giunta, I, f. 113). Latin by Nicolas of Reggio (*c.* 1310). First printed in the Pavia Galen, 1516; re-edited by Guenther, Basel, 1531, and again by Sylvius and Gregoire, Paris, 1538. Sylvius also used it largely in his *Isagoge*. The *De usu partium* provided the foundation of anatomical teaching from medieval times. The substance of the teaching of Sylvius had always been drawn from it.

(c) *De facultatibus naturalibus* (Kühn, II, p. i; Giunta, I, f. 90). Gives a good account of Galen's physiological views. Latin translation by Linacre, London, 1523; reprinted Paris, 1528. Not to be confused with the *De substantia facultatum naturalium* (Kühn, IV, p. 757; Giunta, I, f. 230), a fragment on the ladder of nature and the world-pneuma, translated by Guenther, Paris, 1528.

(d) *De ossibus ad tirones* (Kühn, II, p. 732, Giunta, I, f. 39). The Greek text of this very influential work is in no early collection or edition. Latin translations by Gregoire, with preface by Sylvius, Paris, 1535, and by Ferdinando Balamio, Paris, 1535, are said to have been unknown to Vesalius. This seems unlikely, for Sylvius made use of this work on bones in his lectures and books.

---

* John Caius, *De Libris suis*, London, 1570, pp. 6 *a* and *b*.

(e) *De venarum arteriarumque dissectione* (Kühn, II, p. 779; Giunta, I, f. 55). Latin translation by Antonio Fortolo, Paris, 1526; repeated Basel, 1529. This version was re-edited by Vesalius for the first Giunta Galen, Venice, 1541. Forms the basis of description of vessels in *Tabulae* and *Fabrica*, but see pp. xlvi—lvi.

(f) *De nervorum dissectione* (Kühn, II, p. 831; Giunta, I, f. 53). Latin translation by Antonio Fortolo, Paris, 1526; repeated Basel, 1529. This version was re-edited by Vesalius for the first Giunta Galen, Venice, 1541. Sets forth the classification of the cranial nerves adopted by Vesalius (see note 98).

(g) *De motu musculorum* (Kühn, II, p. 367; Giunta, I, f. 308). This beautiful book is the foundation of the muscular physiology of Vesalius. Latin translation by Nicolo Leoniceno, London, 1522; repeated Paris, 1528. Galen's companion work, *De musculorum dissectione* (Kühn, IV, p. 367; Giunta, I, f. 44), was not printed

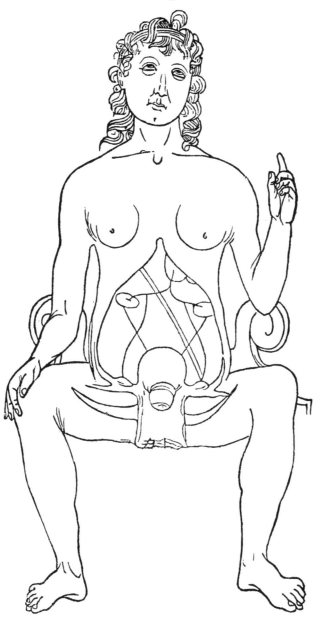

FIG. 14.—From *Fasciculo di Medicinae*, Venice, 1493. The uterus in this figure is the first internal human organ represented from the object in a printed book. The rest of the anatomy is fanciful.

in any **early Greek edition** and did not appear in Latin until the first Giunta, Venice, **1541**. **It was thus** inaccessible to Vesalius in his early Italian years.

(h) *De uteri dissectione* (Kühn, II, p. 887; Giunta, I, f. 108). Also known as *De vulvae dissectione* and *De anatomia matricis*. In all early collected Latin editions of Galen. New translations by G. B. Feliciano, Basel, 1535, by Guenther, Paris, 1536, and by Cornarius, Basel, 1536.

(j) *De temperamentis* (Kühn, I, p. 509, Giunta; I, f. 10). In all collected editions of Galen, both Greek and Latin. A famous Latin version by Thomas Linacre, 1521, was issued by John Siberch, first printer at Cambridge. Contains the lore of humours, diatheses and temperaments that is the basis of all sixteenth century medicine.

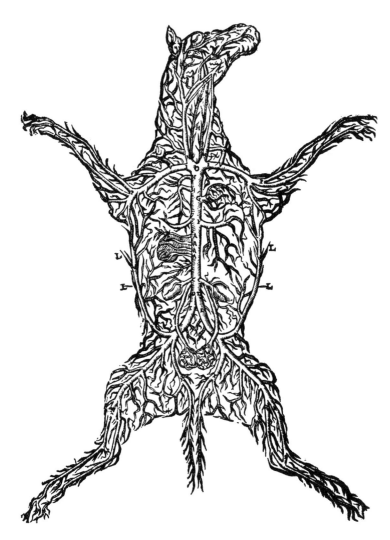

FIG. 15.—From Carlo Ruini, *Dell' anatomia . . . del cavallo*, Bologna, **1598**. **Venous** system of horse, showing continuous vena cava with the right ventricle **as its diverticulum**.

III. Italian Predecessors of Vesalius.

§ 1. *Early Italian Humanist Anatomists.*

When the *Tabulae* appeared, Vesalius had been but eighteen months in Italy and had been very fully occupied. He had therefore not had time to absorb fully the spirit of Italian humanism. He must have read, however, most of the few works on anatomy that had been produced by his Italian predecessors. To understand how these may have affected him we must consider who they were and what they wrote. Apart from Mundinus, whose work he must have examined soon after his arrival, the earliest Italian anatomical work was that of Valla. Vesalius certainly knew it.

Giorgio Valla (1430–99) was a well known humanist and relative of the more famous scholar Laurentius of the same name. He was a polymath who spent much of his life at Venice. There he wrote many works of interest to students of renaissance philosophy and scholarship. Among them is his *De humani corporis partibus*, an interpretation of the anatomical section of the then recently discovered Greek work of Julius Pollux (p. lxviii). Valla is thus the source of several terms which have passed into modern anatomical usage. The first edition seems to have been that produced by the great humanist printer Aldus Manutius at Venice in 1501. It appeared several times, once in conjunction with Guenther's *Institutiones* at Basel in 1536. Vesalius therefore cannot have failed to see it.*

Alexander Benedictus (*c.* 1470–1525) was a medical professor, much interested in anatomy, who had effective command of Greek obtained during residence in the Levant. By about 1490 he was demonstrating anatomy at Padua, where he had a special building for the purpose. His *Anatomice, sive de historia corporis humani*, was printed at Venice in 1502† and several times afterwards. It is poorly arranged and contains no new facts but it heralds Paduan anatomical supremacy. Its significance for us is that it passes the Latino-Arabic texts lightly and goes direct to the Greek of Galen which Benedictus must have studied in manuscripts. After Mundinus his is the first comprehensive work entirely devoted to anatomy, and he is certainly the first anatomist after Valla to draw terms from Pollux. It is likely that Vesalius had read the work of Benedictus before he left Paris, since a student's edition of it had been issued there in 1514 by Henri Estienne.

A contemporary of Benedictus was Antonio Benivieni (*c.* 1450–1502) of Florence. He was a good Greek scholar and friend of many members of the famous Accademia Platonica, among them of Politian (1454–94), first translator both of Plato (1482) and Plotinus (1486). Benivieni left a series of clinical notes, some of which were published in a small volume after his death by his brother (Florence, 1506), himself a distinguished humanist.‡ This collection was re-printed at Paris in 1528 in a volume edited by Guenther. The work records some twenty autopsies. It is the earliest collection of the kind and thus the

---

* Valla also wrote a short *Dialogus Parthenii de sectione humani corporis*, which is hardly anatomical. His *De natura partium animalium* is even less so.

† Fifteenth century editions of Benedictus' work are recorded in several reference books. They are ' ghosts'.

‡ Further cases by Benivieni were printed by F. Puccinotti, *Storia delle medicina*, 3 vols., Leghorn, 1850–6, vol. ii, p. ccxxxv.

first on pathological anatomy. Benivieni seems to have had very little difficulty in obtaining permission to examine bodies. In some cases he even expresses astonishment at the refusal of consent by the relatives. He had a regular technique for post-mortem examinations. It is highly probable that Vesalius was influenced by Benivieni's work which was re-published at Basel in 1529 in a very convenient volume with works edited by Guenther.*

The first writer of the period to make a definite contribution to anatomical knowledge was Gabriele Gerbi (de Zerbis), who died under torture by the Turks in 1506. He taught with success at both Padua and Bologna. His *Anatomiae corporis humani* (Venice, 1502) is drowned in scholastic argument and wretchedly printed but has some points worth notice. Thus it contains what is perhaps the first suggestion for preserving bodies for dissection, namely by the use of ethereal oils. The description of the abdominal organs is fuller than that of earlier writers. The accounts of the main biliary passages, of the uterus and its appendages, and of the urinary bladder and ureters, represent distinct if slight advances. Gerbi has a very high respect for Mundinus and the Arabian writers.

Alessandro Achillini (1463–1512), like Gerbi, taught at both Padua and Bologna. By modern standards his writings are masses of frothy disputation. To glance at them is to raise uneasy reflections that cultured men once treated such windy nonsense with high respect.† He left two anatomical works as gaseously unreadable as his philosophical.‡ They contain, however, the earliest traceable account of the ankle as formed of seven bones, together with a reference to the duct of the submaxillary salivary gland a century and a half before Wharton (1614–73), whose name it usually bears. They also show improvements in the description of parts of the brain, notably of the infundibulum, the anterior ventricles and the fornix. The accounts of parts of the alimentary canal,—duodenum, ileum, colon and caecum,—are advances on those of Mundinus whom Achillini does not hesitate to criticise. He refers to Avicenna as authority for the asymmetry of the renal ' emulgent' veins, two ' branches ' issuing from the left but only one from the right. This is a matter that was later to assume some little significance (p. lxi).

Achillini's writings give some bald directions for dissection, accompanied by a few comments taken from Avicenna. He was best known to his contemporaries as a supporter of the philosophy of Averroes. In 1506, having been driven from Bologna, he became professor of philosophy at Padua, the great Averroan centre. The change from a medical to a philosophical chair was less startling then than it would be now. It is improbable that Vesalius paid any attention to the works of Achillini.

* *Scribonii Largi, De compontione medicamentorum . . . Jo. Ruellius vindicatus; A. Benivenii, De abditis . . . morborum causis; Polybus, De salubri victus ratione . . . Guinterio Johanne Audernaco interprete,* Basel (Cratander), 1529, 12mo.

† There are good accounts of Achillini's writings in Lynn Thorndike, *History of Magic and Experimental Science,* Vol. V, New York, 1941, and in C. F. Mayer's *Bio-bibliography of XVI cent. medical authors,* Washington, 1941.

‡ Achillini's *Anatomiae corporis humani* is included in the same volume as Gerbi's work of 1502. He also wrote *Annotationes in Mundini anatomia.* This was published for him posthumously by his brother Philotheus at Bologna in 1520. There are later editions of both works.

§ 2, *Early Italian Illustrated Anatomies.*

We turn from these semi-medieval texts to early methods of illustrating anatomical observations. The influence of renaissance art did not penetrate medical literature until late and then but slowly. Vesalius was himself the first effective pioneer of the application of art to anatomy. Before the last decade of the fifteenth century there are few anatomical figures in printed books. Such as there are have little scientific or artistic merit.

The earliest printed medical work displaying the influence of renaissance art is the Latin *Fasciculus medicinae*, printed at Venice in 1491. A misunderstanding has attached the ghost-name ' Ketham ' to this and to a number of similar later volumes. The Latin *Fasciculus* of 1491 is a miscellaneous collection of popular Latin medical tracts, mostly of the fourteenth century. It happens to contain, however, several fine figures by a good artist. These have hardly any anatomical interest.* Two years later the same printer produced an Italian *Fasciculo di medicina.*† This beautiful book is in no sense a second edition of the Latin *Fasciculus medicinae* of 1491. It is a different work, in different format, different type, different language and with different contents and some different figures. The name Ketham is appropriate to neither. It is best to distinguish these two works, and similar works which follow them, by the Latin or Italian title, with added dates as above. ‡

The frontispiece of the *Fasciculo di medicina*, 1493, presents a Paduan professor § lecturing, with the usual medieval medical books around him (Fig. 13). The longest text in the collection of tracts that make up this *Fasciculo* is the *Anothomia* of Mundinus, illustrated by a magnificent dissection scene (Fig. 1). This is the best early presentation of an academic anatomy. Another item in the volume is a fine figure in a traditional pose exhibiting the structure of the female organs. The anatomy of this figure is erroneous and traditional, but the uterus itself is drawn with at least a memory of the object (Fig. 14). It is the earliest instance in a printed book of naturalistic representation of an internal organ. It is the work of an excellent draughtsman. A resemblance to certain drawings by Leonardo has been suggested.

Illustrated anatomical textbooks do not appear till the third decade of the sixteenth century.∥ The first are those of Giacopo Berengario da Carpi (died

---

* The *Fasciculus medicinae*, falsely ascribed to the ' ghost ' Ketham, has been reproduced in *Fasciculus medicinae of Johannes Ketham Alemannus*; *Facsimile of the First (Venetian) Edition of 1541, with introduction by K. Sudhoff, translated and adapted by Charles Singer*, large folio, Milan, 1924.

† *The Fasciculo di Medicina Venice, 1493, with introduction and translation by Charles Singer*, 2 vols. quarto, Florence, 1925.

‡ The confusion is increased by the fact that several later Latin 'editions' are taken partly from the Italian *Fasciculo* of 1493. These editions, also, unfortunately, called Ketham, have some altered figures of their Italian original and texts which vary from edition to edition. For our purpose only the *Fasciculus* of 1491 and the *Fasciculo* of 1493 are significant.

§ In the picture he is given the name Pietro de Montagnana. There were several bearing the name Montagnana in fifteenth-century Padua. Pietro, however, has not been identified.

∥ The coarse, ugly and fanciful diagrams in the works of Brunschwig (Augsburg, 1497), Peyligk (Leipzig, 1499), Hundt (Leipzig, 1500) and Reisch (Freiburg, 1502) and the noble drawings in the *Fasciculo di medicina*, 1493, can none of them be said truly to illustrate anatomical texts.

1550). * He claimed to have opened several hundred bodies (*ego quamplurima centena cadaverum secuerim*). Clearly most of these must have been autopsies rather than anatomisations, but he published several works which display much first-hand anatomical knowledge. Though the first of these is presented as a commentary on Mundinus (1521), all are original contributions. Berengar criticises the work on which he professes to comment. For example he denies the existence

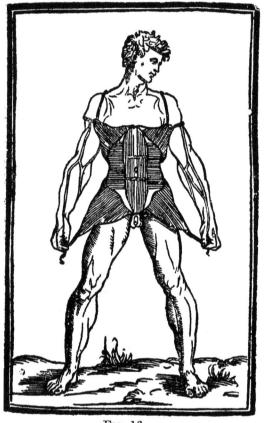 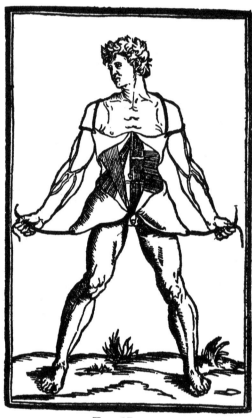

FIG. 16.   FIG. 17.

FIGS. 16 and 17.—The abdominal muscles, from Jacob Berengar of Carpi, *Isagogae breves*, Bologna, 1522 and 1523.

of the *rete mirabile* below the brain, though the description of it by Mundinus is based on Galen. It is the earliest anatomical treatise with figures illustrating the text. These represent (*a*) the abdominal muscles (Figs. 16 and 17) in six plates of some anatomical and artistic merit; (*b*) the superficial veins of the extremities, which are important for the contemporary practice of bleeding, in three poorly executed plates; (*c*) the uterus and adnexa in three wretched plates ; (*d*) two plates of considerable beauty illustrating surface markings. Vesalius had probably read Berengar before 1538.

* This name has caused confusion. He was born *c.* 1470 at Carpi, in the Ferrarese territory, son of a physician, Faustino Barigazzi. In the Bologna rotuli 1502–22 he figures as Jacobus de Carpo. He was proud of his birthplace, falsely deriving it from Greek *karpos* = fruit. In 1500 his father fell foul of the house of Ferrara. Giacopo then dropped the name Barigazzi and later took the name of his wife's family, Berengario. In his three anatomical works, *Anathomia Mundini noviter impressa ac castigata*, Bologna, 1514, *Commentaria cum amplissimis additionibus super anatomia Mundini una cum textu ciusedm in pristinum et verum nitorem redacto*, Bologna, 1521, and *Isagogae breves . . . in anatomiam humani corporis a communi medicorum academie usitatam*, Bologna, 1522, as well as in his surgical work *De calvariae sive cranii fractura*, Bologna, 1518, he is called Jacopus Berengarius Carpensis. In the *Galeni . . . Libri anatomici*, Bologna, 1529, he is simply Jacobus Carpus. Legal documents still call him Barigazzi in 1506. See G. Martinotti, ' Il testamento di Magistro Jacopo Barigazzi, o Berengario de Carpi ' in *Rivista di Storia delle Scienze*, XIV, Siena, 1923, and Vittorio Putti, *Berengario da Carpi*, Bologna, 1937.

A move toward fuller illustration is found in a work by Berengar of 1522, the title of which may be translated *A Short but very clear and fruitful introduction to the anatomy of the human body, published by request of his students.* * It has figures from his *Commentaria* on Mundinus, together with new figures, among them of the heart and its valves (Fig. 39), and of the uterus with attached tubes and ovaries. Rude though these are by our standards, they establish the book as the first anatomy illustrated in the modern sense. Moreover the book initiates one excellent and fatigue-saving practice, unfortunately abandoned by Vesalius and his followers. Berengar inscribes the name of the organ on its figure and is not content with mere lettering. For introducing this simple and effective device the weary student should continually bless his name (Fig. 39).

Berengar is the first to describe and figure the axis adequately, the first to produce intelligible figures of the structure of the heart, the first in modern times to distinguish the chiliferous vessels as distinct from the veins, the first to describe the lachrymal duct, the first to show experimentally that no branches of the renal veins open into the excretory ducts, the first to describe the vermiform appendix, the first to see the arytenoids as separate cartilages, the first to recognise the larger proportional size of bony thorax in the male and of pelvis in the female, the first to give a clear account of the thymus gland. He knows something of the action of the cardiac valves. His description of the brain is an advance on Mundinus, recognising the general form of the ventricles, the formation of the chorioid plexus from arteries and veins, the pineal gland and the relations of the fourth ventricle,

Berengar was a popular and effective teacher. His latinity has been judged as excessively debased. In view of the peculiar unfitness of renaissance Latin for scientific purposes, Berengar's avoidance of its maddening involutions and portentous adjectival clauses appears as a virtue, a sign of life rather than of decay. Would that Vesalius had been as direct and as sensible in his language. Berengar is responsible for the term *Vas deferens*, which has been criticised as a misnomer by purists. † There are many such imperfect derivations in the work of Vesalius. Modern anatomical nomenclature is full of them. Vesalius took the term vas deferens from Berengar.

The so-called 'fugitive anatomical sheets' were a peculiar development of anatomy in the first half of the sixteenth century. On them were printed figures showing the bare outlines of anatomy. They were used by students of medicine, who then, as now, were resolved not to burden their memories with superfluous details. They were probably technically inferior even to the rude illustrations in the anatomical text-books of the time. ‡ There are, however, certain fugitive sheets that are more significant. They were intended not for medical, but for art students, and show the interest in Anatomy in that quarter from which the real reform of Anatomy had come. One such sheet by Rosso de' Rossi (1496–1541, see p. xix) in copper-plate is almost exactly contemporary with the first work of Vesalius. §

* *Isagogae breves perlucidae et uberrimae in anatomiam corporis humani, ad suorum scolasticorum, preces in lucem editae, cum aliquot figuris anatomicis*, Bologna, 1522 ; second edition 1523. Editions described in the bibliographies as 1514, 1518 and 1521 are non-existent. Their dates are due to confusion with other works.

† The prefix *de* indicates *down*, whereas the vas leads up !

‡ Le Roy Crummer, ' Early Anatomical Fugitive Sheets ', in *Annals of Medical History*, 1923 and 1925.

§ L. Choulant and M. Frank, *History and Bibliography of Anatomic Illustration*, New York, 1945, pp. 30 and 113–15.

f

§3. *Massa (c.* 1480–1569).

A contemporary of Berengar was the Venetian practitioner Nicolo Massa (*c.* 1480–1569), who was educated at Padua. This remarkable anatomist is seldom noticed, for he had no academic position and his book no illustrations. Nevertheless he is important in the history of anatomy, and specially so for the estimation of Vesalius. The *Introduction to Anatomy by Nicolo Massa, the Venetian, Doctor of Arts and of Medicine, a Book of Dissection of the Human Body, wherein are described many Organs disregarded by both Ancients and Moderns as will be clear to Studious Readers,* appeared in 1536.* It is sensible, modest, practical, simply written, based on first-hand knowledge, and the best short account of the structure of the body to its time. Its dedication to the Farnese Pope (Paul III, 1534–69) significantly emphasises the urgent need of dissection for medical men.

Massa indicates that he had examined many bodies in the presence of other physicians. Certain of his autopsies were held at the hospital of St. John and St. Paul at Venice.† It is evident that he had the public confidence and that post-mortem facilities were readily granted him by the relations of his patients. He records that one autopsy was actually suggested by the son of a patient whose name he gives. Much of his knowledge was obtained from the bodies of still-born infants. The outlook of his book contrasts with that of the *Tabulae* of Vesalius, which was published at Venice just two years later.

Massa was some thirty years older than Vesalius. His anatomical activity began when Vesalius was an infant. His book summarises the long experience of a busy, trusted, elderly practitioner, writing primarily for colleagues in a similar position to himself. The *Tabulae,* on the other hand, is an early academic exercise by a young newly appointed professor seeking to expound to students the elementary bases of his subject. Massa has far wider experience both of human beings and of human bodies than Vesalius, and he has a correspondingly less detached and more human approach. He is no less aware than Vesalius of differences between apes and man, and as constantly quotes Galen, from whom he does not hesitate to differ. He often acknowledges his debt to the 'moderns', notably Mundinus, Gerbi and Berengar, and is pleasantly free from the tiring abusiveness of the humanist writers of the day. His nomenclature is perhaps rather more antiquated than that of the *Tabulae.*

Massa is the first to use the term *panniculus carnosus* (=**platysma myoides** of Galen). He observed that this muscular coat in man is thinner, less developed and much more local than in animals. Nevertheless he greatly exaggerates its extent and importance. In a routine account of the abdominal wall he describes the inguinal canal and compares the tendinous intersections of the *rectus abdominis* to that of the digastric muscle. He gives a fair account of the anatomy of hernias which he had often dissected. His description of the intestinal canal is accurate and includes the appendix, which he thinks tends to disappear in adults.‡ He has observed the great variability in size of the spleen and has often seen it extending into the lower abdomen.

Massa is dissatisfied with the conventional medieval scheme of a liver with five equal lobes (such as is typically displayed by the *Tabulae*), and he emphasizes

---

* *Anatomiae liber introductorius,* Venice, 1536. We have worked with the reprint of 1559.

† The Church of that name, a very fine fifteenth-century building, still stands.

‡ This idea may be taken from the ape in which the appendix is absent in the adult.

the superficial division of the viscus into two main lobes. He has an interesting passage, the first of its kind, on the internal distribution of the hepatic vessels. " The *vena porta* ", he says, " divides in the substance of the liver into five veins which traverse the five lobes and then ramify in many branches in the convexity, whence emerges the *vena cava* (Fig. 58). But you will note that these minute branches are united with the branches of the *vena cava*. This you will see better if you macerate the liver for some days and then boil it so thoroughly that the flesh can be separated easily from the vessels. You will then perceive the substance of the veins to be interwoven, as it were, into a sort of network." This is followed by an account of the 'branches' of the *vena cava*, fuller and more accurate than that of the *Tabulae*, and more than equal in clarity to the text of the *Fabrica*.

In describing the kidneys Massa proves experimentally by injection that the cavity of the renal veins is not continuous with that of the sinus of the kidney. This was a real physiological advance since the kidneys were regarded by the Galenists as literally filters for straining off the urine from the blood. Thus Mundinus had written: "The blood is filtered in the hollow of the kidneys. The way of it you can see by cutting the kidney longitudinally, when a membrane or thin covering (= wall of sinus) becomes visible. This is the wall of the *vena emulgens* (= renal vein), thinned to a fine sieve-like surface. Through the pores of this there passes urine but not blood". But Massa writes : " As to the structure (*substantia*) of the kidneys and their inward parts. Raise the kidney, peeling it carefully so as to break no vessel, and you can ascertain beautifully whether the foramina of the sieve (= calyces of the sinus) are perforate. To ascertain this, place a quill in the emulgent (= renal) vein and blow through it. You will see the kidney swell up, but no spirit or air will come to the *pori uritides* (= ureters) or emerge through them." A good physiological experiment repeating Berengar's (p. xxxiii).

Massa records the differences in origin of the seminal arteries and veins on the two sides and it may well be that he had made this observation before Vesalius was born. In the very year of publication of Massa's work Guenther unaccountably ascribed this discovery to Vesalius, though Massa, Berengar and Mundinus were all familiar with the facts (see p. lxi). Massa gives a short account of the prostate and is the first anatomist to record it. His descriptions of the uterus and female genitalia surpass those of the *Tabulae* and *Fabrica*, and are based on much more material than was available to Vesalius, but the account of these organs is one of the weakest sections of the *Fabrica*. In his chapter on the thorax, Massa's account of the heart contains a number of points of interest, notably that the 'third ventricle' of the medieval anatomists cannot ordinarily be seen and that the interventricular septum is not porous but dense, hard and solid. He concludes that the third ventricle is an error. He describes the *rete mirabile*, though he admits that it is sometimes very difficult to see, since with the lapse of time after death it is liable to become inconspicuous! Yet it is evident that there were sceptics among his audience for he exclaims that " some dare to say that this *rete* is a figment of Galen . . . . but I have myself often seen the rete, and I have demonstrated it to the bystanders so that no one could possibly deny it, though sometimes I have found it very small ". To make the demonstration easier he suggests the ligation of the carotids and jugulars before the brain be opened (see p. xliii).

It is impossible to believe that Vesalius had not read Massa, and it is difficult entirely to acquit Vesalius of concealing his debt to a man so near to him in time, place and interest. Certainly Vesalius passed anatomically far beyond Massa, though not in the *Tabulae*. But before any anatomical discovery be ascribed to Vesalius it would be safer to consult the work of Massa.

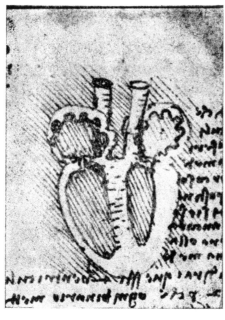

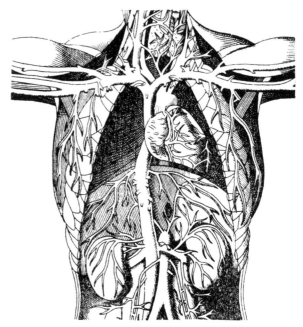

FIG. 18.   FIG. 19.

FIG. 18.—Cavities of the heart, by Leonardo da Vinci (early sixteenth century). *Quaderni d'anatomia*, folio 3, recto. Leonardo writes : "The ventricles are separated by a porous wall through which the blood of the right ventricle penetrates into the left."

FIG. 19.—The great vessels from Eustachius, *c.* 1550 (edition of Albinus, Leyden, 1744). Innominate artery, left carotid and left subclavian shown arising separately from the aortic arch. The vena cava superior and its great tributaries are less correct and have some simian features. Right kidney and renal vein are above left.

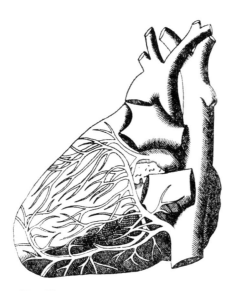

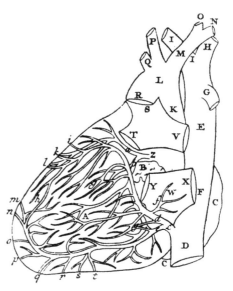

FIG. 20.—From Eustachius, *Tabulae anatomicae* (edition of B. S. Albinus, Leyden, 1744). Heart and great vessels from behind. Vena cava inferior and superior shown as continuous. Branches of the aortic arch are almost correct. Junction of the two innominate veins a little higher than the top of the aortic arch. B, left auricle ; C, right auricle ; D, E, F, vena cava ; G, azygos ; K, L, R, aorta ; S, T, V, pulmonary artery ; W, X, Y, pulmonary vein ; Z, left coronary artery; c, left coronary vein.

IV. Galenic Physiology and its Latin Presentation.

§ 1. *Basic Principle of Galenic Physiology*.

Standing behind all anatomical teaching before the seventeenth century is Galen's physiological system. It is set forth or implied in the various texts that were printed in the early collected editions of Galen's works. Until the boyhood of Vesalius however Galenic teaching reached the student of medicine mainly through Avicenna and other Arabs and Arabists. By the time Vesalius began study at Paris (1533) the anatomical works of Galen had been translated into Latin by Guenther and others (pp. xxv–vii). The physiological views of the Middle Ages, being themselves inherited from Galen through Avicenna, could, with but slight adjustment, be made to conform to the newly recovered texts. The main need was re-expression in the humanist dialect of the time. In this none was more expert than Giovanni Battista de Monte (Montanus, 1498–1551), the most popular medical teacher in North Italy. He was in intimate contact with Vesalius, and in 1538 was called to a chair at Padua. In the *Fabrica* Vesalius expresses admiration for him. In the *Tabulae* Vesalius was in effect putting into graphic form the physiological teaching of this man.*

Even with such a teacher as Montanus it must be admitted that the art of scientific exposition was as yet little developed. Vesalius himself had a share in perfecting the technique for this. It is therefore very unfortunate that the clarity of his figures is obscured by his language. The account which follows is an attempt to set out the old physiology in the modern manner. This is necessary for Galenic physiology is a lantern in the fog of verbosity that enshrouds medical discussion of the day.

The first need for understanding the old physiology is to shed not only the idea of a circulation but also the conception that the passage of the blood through the vessels is a very active process. Blood, it was held, flows to and from the liver through the veins, of which the chief are (*a*) the portal vein, (*b*) the vena cava and (*c*) a special diverticulum of the vena cava that we call the right ventricle. The left ventricle sends out only a specially modified fraction of the blood. The atria have but a subordinate function ; they are mere safety-outlets for excess of blood from the ventricles. With these points in mind we turn to consider the system as a whole.

The basic principle of life in the Galenic physiology is a *spirit, anima* or *pneuma*, drawn from the general world-soul in the act of respiration. This pagan Stoic principle is, strictly interpreted, inconsistent with Christian, Moslem, or Jewish doctrine, an awkward fact omitted or slurred or misrepresented or misunderstood in medieval writings. Thus the medieval physiological system is basically unintelligible. With the advent in the sixteenth century of translations of the more philosophic works of Galen, and in the semi-pagan atmosphere of renaissance humanism, Galenic physiology could be presented in a form nearer to the original. Nevertheless the ' world-soul ' of Galen and the Stoics (as well as of the Christian and Jewish Averroists, who were strong in Padua) could never be made to fit the picture of the world demanded by the Church. It is a very old divergence. In the Gospel according to St. John there is a familiar passage which

---

* A full expression of the physiological views of Montanus is in the work of his student, Valentine of Lublin. It is in effect a notebook of his master's lectures : *J. B. Montani . . . in priman Fen Libri primi Canonis Avicennae Explanatis*, Venice, 1554.

approaches and yet avoids this very antithesis between Stoic and Christian : "That which is born of the Spirit (*pneuma*) is spirit (*pneuma*) . . . The wind (*pneuma*) bloweth (*pnei*) where it listeth . . . and thou knoweth not whence it cometh : so is everyone that is born of the Spirit (*pneuma*)."—John iii.

This world-pneuma or spirit of Galen enters the body through the trachea, the 'rough artery' (*arteria aspera* of medieval notation). Thus the pneuma passed to the lung and thence through the vein-like artery or *arteria venalis* of medieval writers (our pulmonary vein), to the left ventricle. Here it undergoes a change which we consider below.

§ 2. *The Galenic Physiological Scheme.*

Food traversing the alimentary tract is absorbed as 'chyle' from the intestine, collected by mesenteric veins into the portal vessel by it and conveyed to the liver. That organ has power to elaborate the chyle into venous blood and to imbue it with a spirit or pneuma which is innate in all living substance, the *natural spirit* (*spiritus naturalis* of the medievals) and is ultimately derived from the world spirit. Charged with natural spirit, but also with the nutritive material derived from the food, venous blood is distributed to the body by the liver through the vena cava, which runs the length of the body. This vena cava arises from the liver much as the aorta arises from the heart, that is to say from the left ventricle. The veins which branch out from the vena cava carry nourishment and natural spirit to all parts of the body. *Jecur venarum principium*, 'the liver, source of the veins', is first of the keys to Galenic physiology. The blood ebbs and flows continuously in the veins.

There is one great branch of the *vena cava* which has a special significance. This is the diverticulum which enters the right side of the heart. It is, in effect, the cavity of the right ventricle itself, from which the right atrium is a safety valve for any overflow.

The blood in the system of vessels derived from the *vena cava* ebbs and flows in the body, carrying nourishment and 'natural spirit' to all parts. But for the blood that thus reaches the right ventricle two fates are possible :

(*a*) The greater part remains awhile in the ventricle, parting there with the impurities that it has brought from the organs. These are carried off by the artery-like vein (our pulmonary artery) to the lung and thence exhaled to the outer air. These fumes give its poisonous and suffocating character to the breath. Having parted with its vapours, the venous blood from the right ventricle ebbs back again into the general venous system.

(*b*) A small fraction of the venous blood that enters the right ventricle behaves, however, in another way. Drops of this venous blood, charged still with the natural spirit derived from the liver, are always trickling through minute (imaginary) channels in the septum between the ventricles (**Figs. 18 and 21**).* This fraction of the venous blood thus reaches the left ventricle. There it encounters the pneuma from the outer world that has come to that cavity through the *venal artery* (our pulmonary vein) from the lung, whither it is brought through the trachea. By the action of this pneuma or air the blood in the left ventricle, or rather the natural spirit in that blood, is elaborated into a higher form of spirit, the *vital spirit* (the *spiritus vitalis* of the medievals). This is distributed by the arterial system to

---

* These minute channels have, by a misunderstanding perhaps derived from Aristotle, become represented as a 'third ventricle' within the septum in the *Canon* of Avicenna and works derived therefrom.

various parts of the body.   *Cor vitalis facultatis fomes et arteriarum principium*
' the heart, nurse of the vital faculty and source of the arteries ', is the second
key to the Galenic scheme.

Blood containing vital spirit ascends to the brain via the ' soporal ' or ' apoplectic '
arteries (our carotids).   These vessels divide there into minute channels to form
the *rete mirabile* (Figs. 21, 23, 26 and Tabula III).   In this ' marvellous network '
the blood is minutely divided for the third time.   Thereby the vital spirit which it
contains is transformed into a third and yet higher type, the *animal spirit* (*spiritus
animalis* of the medievals), a yet more subtle substance.   *Rete mirabile in quo*

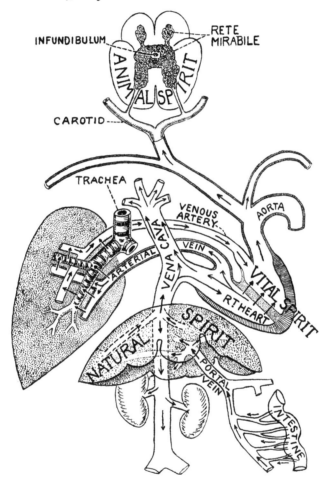

Fig. 21.—Diagram illustrating Galen's physiological scheme.

spiritus vitalis ad animalem preparetur, ' the wonderful net wherein vital is
elaborated into animal spirit ', is the third key to the Galenic scheme.   This third
essence is distributed to the body by the nerves.   These contain minute channels
through which the animal spirit goes to all the organs, and especially to the muscles.

According to the pagan Galen, writing under the influence of the Stoic philosophy, the three fundamental faculties, the natural, the vital, and the animal,
which correspond to these spirits and bring into action the corresponding functions
of the body, originate as an expression of the primal force or world-pneuma.
These spirits are well recognised as entities in medieval physiology though
necessarily misconstrued.

This highly ingenious physiology is not derived from an investigation of
human anatomy.   It is fanciful and, in so far as it is related to actual structures,
they are those of animals and not of men.   Thus in the human brain there is no

rete mirabile, though such an organ is found in the calf.   Within the human thorax
the course of the inferior vena cava is extremely brief, though in monkeys, as in
all domestic mammals, there is a considerable stretch of the vessel between the
heart and diaphragm.*   Galen states that he dissected many kinds of animals—
sheep, oxen, asses, mules, swine, lions, wolves, dogs, lynxes, bears, weasels, apes,
camels, mice, serpents, fish, birds and once an elephant.   He specifically states
that he did not dissect human bodies, either of enemies killed in war or of exposed
children,† but he does not say that he never dissected a foetus.   His anatomy is
derived mostly from apes, supplemented for special purposes from cattle and swine.

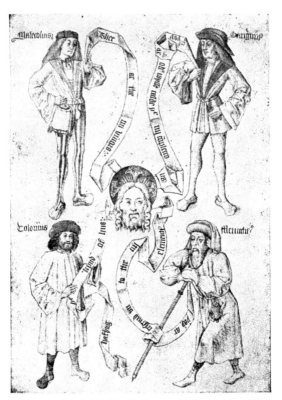

FIG. 22.—The four temperaments, from the Guild Book of the Barber Surgeons of York of about 1500,
    now in the British Museum.   Melancholy, Choleric, Sanguine and Phlegmatic types are
    shown.   On the scroll is written : "Ther ar the iiij umors.   That ar oderwysse calde the iiij
    complecionis, that ar rescevid unto the iiij elementis, Hafying the kynd [=nature] of humors."

Some of Galen's anatomical errors were due to his attributing to one creature
structures found in another.   It was among the achievements of Vesalius
that he demonstrated this.   But it cannot be maintained that he was the first to
discover it, nor was he able to improve on Galen's physiological system.   Moreover,
much of the anatomy of Vesalius is itself based on that of animals.

With the Galenic theory of the action of the vascular system there was combined the medieval version of the doctrine of the four humours, *Blood, Phlegm,
Black Bile* (Melancholy) and *Yellow Bile.*   The four Aristotelian elements, Earth,
Air, Fire and Water, were supposed to be the origin of these, though Aristotle does
not, in fact, mention the humours.   The older Hippocratic writings, on the other
hand, while they discuss the four humours, make no mention of the four elements.

---

* The extreme reduction of intra-thoracic course of inferior vena cava in man is associated with
absence of ' azygos ' lobe of the right lung, which is well developed in the ape and domestic mammals.
    † Galen, *De anatomicis administrationibus,* III, Ch. 5 ; Kühn, II, pp. 385–86.

Each of the four humours was associated by the medieval anatomists with a special organ. There are thus four 'principal organs'. Blood arises from the *liver*, Phlegm from the *brain*, Yellow Bile or Choler from the *gall-bladder*—which is thus of peculiar importance—and Black Bile, or Melancholy, from the *spleen*.

Each of the four humours must be evacuated when in excess and each has its own 'emunctory' or cleansing place. Blood is evacuated through the nose or mouth or by the menstrual process. Phlegm (=*pituita*) emerges from the brain (hence our 'pituitary body') through the cribriform plate into the nostrils. Yellow bile is evacuated from the gall-bladder through the common bile-duct into the intestine. The evacuation of black bile from the spleen provided a difficulty. Some medieval writers assumed a special duct leading from the spleen to the stomach, some assumed that the black bile was carried from spleen to stomach by blood-vessels. Vesalius had his own special views on this last point (note 19).

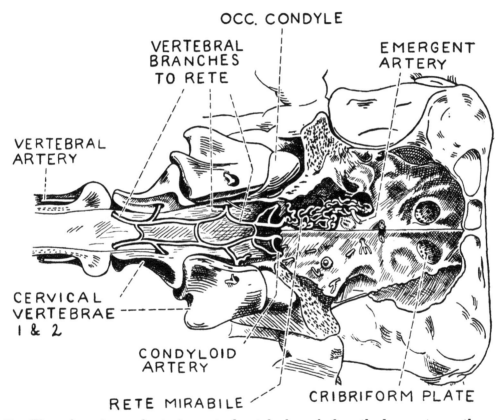

FIG. 23.—Floor of cranium and anterior part of vertebral canal of ox, the dura mater partly removed on one side to expose the *rete mirabile*, modified from A. G. T. Leisering's *Atlas der Anatomie . . . . der Hausthiere*, Leipzig, 1866.

Lastly the four humours determine the human 'dispositions'. In health the four make a perfect mingling, or 'complexio' or 'temperament'. In practice, however, one or other is liable to be in excess, and the complexion or temperament is liable to accord to one of four well known types—the Sanguine, the Phlegmatic, the Choleric or the Melancholic (Fig. 22). Human dispositions are still recognised as corresponding to these types, so that the words retain to this day a residual element of their medieval meaning. The whole doctrine of the humours has become deeply implicated in literature.*

* The subject is entertainingly traced by P. A. Robin, *The Old Physiology in English Literature*, London, 1911.

V. Certain Anatomical Elements in the ' Tabulae '.

§1. *Reliance of Vesalius on Animal Anatomy.*

In the *Fabrica*, Vesalius violently and repeatedly denounces Galen for describing animal structures while professing to give an account of the human frame. Yet it is from animals that Vesalius drew anatomical conceptions not only for the *Tabulae* but also for the *Fabrica* itself, as we shall presently show. But neither his censoriousness nor his errors should be too readily or too severely condemned. They must be judged in their own peculiar atmosphere.

Abusiveness and unrestrained criticism were almost normal to the humanist literary tradition of his day. These things render many writings of the period quite unendurable to modern readers. Not a few of the most distinguished renaissance scholars produced their greatest works in moods which seem to us now both childishly irritable and flatulently abusive. Difference of opinion on a purely technical point is often expressed in language of the coarsest vituperation, unprintable in a modern vernacular. Denunciations by Vesalius must therefore be read with the detestable literary manners of his age in mind. He is far from being the most hypercritical of humanist writers. Among them he may fairly be described as a man moderate in tone.

Nor was it reprehensible in Vesalius to draw on the anatomy of animals. Indeed it would have been remembered to him for righteousness had he not protested too much and too often that he alone described only the parts of man. It was the very reliance on comparative methods that gave a special distinction to the grand Paduan anatomical school that he adorned. Without an appeal to animal bodies, what would have been the achievement of Fabricius and Coiter, Harvey and Casserius, and a dozen others ? And when the Paduan professors abandoned the comparative method, did not the greatness of their school fade with it ?

Certainly Vesalius must have dissected far more animals than men. For investigating many anatomical points animal bodies are better adapted than human. Their dissection could be conducted in his own chamber. A glance at the title-page of the *Fabrica* arouses the reflection that privacy must have been the pressing need of this noisy, bustling, exhibitionist genius. His anatomical vision was perforce a patchwork construction. Its basis was the dissection of animals and still-born children, corrected and checked at rare intervals from the corpses of adults, almost all of them males. That is the best that could have been. The marvel is that the vision of the *Fabrica* became at once so complete and so unified. And remember that it was a vision, for Vesalius had ever the complete and living human form in view.

Nevertheless there are many details, both in the *Tabulae* and in the *Fabrica*, that have not been corrected, or have not been fully corrected, by reference to human bodies. It is not remarkable that these should be more evident in the work of 1538 than in that of 1543. Some features drawn from animals in the earlier work are corrected in the later. We shall not here exhaust all the non-human details, even of the *Tabulae*.

§ 2. *Rete mirabile.*

The method by which a mental process initiates and controls muscular action was as mysterious to the ancient as it is to the modern philosopher. But the ancient physiologist, like the modern, gave much attention to the actual mechanism of the nervous system. The elaborate descriptions of the physiologist were, as they still are, apt to mislead the unphilosophic into the naïve error that an exact relation between mind and body had been displayed. In the Galenic system the *pneuma* of the outer world was supposed to come by the trachea to the heart, where some of it met the *natural spirit* which was thus changed into *vital spirit*. This was then sent to the brain, where in the complex *rete mirabile* it became *animal spirit*, to be distributed to the body via the nerves. Thus the rete was essential to the working of the system (p. xxxix).

The actual *rete mirabile* of comparative anatomy is an elaborate network of vessels into which the internal carotid divides at the base of the brain. It is formed just before the chorioid artery is given off and is a very conspicuous feature in Ungulata. There is a much less developed *rete mirabile* in the Carnivora, but no trace of it in man, monkey or rodent. Galen derived his conceptions of the rete and its nature from Herophilus. He studied it specially in the ox and sheep (Fig. 23). The actual function of the *rete mirabile* is quite unknown.

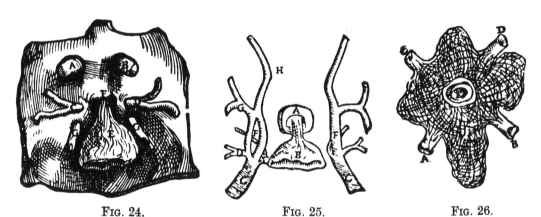

FIG. 24. FIG. 25. FIG. 26.

FIG. 24.—From *Fabrica*, p. 60. Vesalius writes : "Part of the cranial cavity formed by the *os cuneus* [*i.e.* sphenoid bone] still covered by dura mater. A, B, parts of visual nerves ; C, D, the arteries which, having penetrated the dura, are distributed partly to the pia mater and partly to the ventricles [*i.e.* internal carotids] ; E, the basin (*pelvis*) which receives the phlegm dripping down from the third ventricle [*i.e.* the *infundibulum* torn from the brain with, perhaps, part of the *tuber cinereum* attached] ; F, aperture through which the end of this funnel-like basin reaches the gland [*i.e.* the pituitary] which receives the phlegm of the brain [*i.e.* the opening in the *diaphragma sellae*] ; G, second pair of nerves of the brain [*i.e.* *oculo-motor*]."

FIG. 25.—From *Fabrica*, p. 621. "We here depict the [pituitary] gland laid bare, with the basin or funnel [*i.e.* *infundibulum*], which conducts the phlegm down to it, hanging flaccid. Laterally we indicate, just as they have appeared to us in dissection, parts of the soporal arteries [*i.e.* carotids] which are alleged to form the reticular plexis. Since they are variable, we have drawn variants. A, the [pituitary] gland ; B, the basin or funnel ; C, C, the arteries passing obliquely forward in their proper channels in the skull."

FIG. 26.—From *Fabrica*, p. 621. "We here represent the plexus [*i.e.* *rete mirabile*] falsely, but according to the description of Galen's *De usu partium*. A and B represent the arteries at the base of the brain which are supposed to break up into the *plexus mirabilis* ; C and D are the vessels into which the branches of that plexus are supposed to be united. They correspond in size to A and B ; E is the [pituitary] gland."

The emphasis laid by Galen on the *rete* would in itself betray the fact that he relied on animals for his anatomical accounts. It further shows how strongly his anatomy was controlled by his physiological conceptions, as indeed was that of Vesalius. During the Middle Ages the existence of the *rete* in man was not doubted. Mundinus (1316) carefully describes it but adds that it can be seen well only in bodies very recently dead. The same view is expressed by Estienne (1539). The more independent Berengar (1522) admits that he could not find it, though Massa (1536), a practical and experienced anatomist, had no doubt of its existence (pp. xxxv–vi).

In Tabula III Vesalius has drawn, to quote his own phrase, ' a reticular plexus at the base of the brain, the *rete mirabile*, wherein the vital is elaborated into the animal spirit '. It surrounds the infundibulum completely. From this circle of the rete two pear-shaped processes pass into the brain. These are the

FIG. 27.—*Macaca mulatta*, from frontispiece   FIG. 28.—*Macaca inua*, the Barbary ape, from
            of *Fabrica*.   R. Lydekker.

' choriform plexuses formed from arteries and veins in the cerebral ventricles '. Here in Tabula III is the first attempt in a printed book to portray the *rete*, though there are some earlier parallels in MSS. In the *Tabulae* Vesalius expresses no doubt of the existence and form of the rete. However, by the time of the appearance of the *Fabrica* he perceived and acknowledged that he had been describing what was not there. He writes : " I cannot sufficiently marvel at my own stupidity ; I who have so laboured in my love for Galen that I have never demonstrated the human head without that of a lamb or ox, to show in the latter what I could not in the former, lest forsooth I should fail to display that universally familiar plexus. For in no way do the carotids form a *plexus reticularis* (in man) as Galen alleges." (*Fabrica*, p. 642.)

A franker recantation is hardly possible. It is however but just to remember the doubts of Berengar twenty years earlier (p. xxxii). Moreover, in his description of the parts in the *Fabrica* (Figs. 24 and 25) he makes certain errors of his own.

Some of these, which we have discussed elsewhere, are based on certain modifications which he himself introduced into Galenic physiology.*

§3. *Position and Form of Heart.*

The heart drawn in Tabula III is not the human organ.   We believe it to be that of an ape, probably *Macaca mulatta*, the rhesus monkey, the creature figured among the animals for dissection on the title-page of the *Fabrica* (Fig. 27).

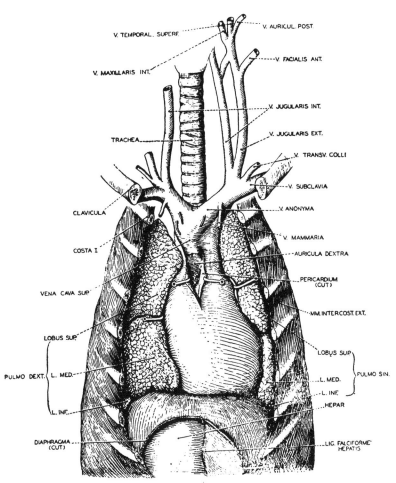

Fig. 29.—Rhesus monkey :   contents of thorax.   From H. Linebach in C. S. Hartman and W. L. Straus, *Anatomy of Rhesus Monkey*, Baltimore, 1933.

The heart of the rhesus differs considerably from that of man (Figs. 29 and 44). Viewed from the front it is relatively nearer the mid line and its axis more vertical.   This arrangement, with a more elongated oval outline, is a natural accompaniment of a thorax more laterally compressed than the human.   Roughly speaking, the ape's heart is longitudinal, man's transverse.   Moreover, the lower part of the wall of the right ventricle in the ape often projects slightly, giving then almost the appearance of a double apex.   Further, there are two anterior branches of the left coronary artery in the ape instead of the one normal in man. All these points come out in Tabula III which, as we shall show, was drawn with an ape in mind.

In the *Fabrica* Vesalius describes the contents of the thorax in the sixth of his seven books.   The heart and lungs there represented are certainly of human

* Charles Singer in *Journal of the History of Medicine*, I, 1946.

subjects (Figs. 32 and 33). Yet throughout *Fabrica* and *Epitome*, as in the *Tabulae*, the great vessels are consistently represented with several non-human traits. Some of these features are derived from the rhesus monkey with, perhaps, hints drawn from the human foetus; others are derived from ungulates. All show the influence of Galen.

§4. *Branches of Aortic Arch.*

The aorta in Tabula III is unlike the human type. The most obvious difference is that the arch gives off not three but only two branches. The first of these is a large *truncus communis* which gives rise to (*a*) the innominate and (*b*) right subclavian. The second is the left subclavian. In the *Fabrica* and *Epitome* (Figs. 30, 31, 32 and 43) Vesalius consistently adheres to this presentation, even in Book VI where, portraying the thorax of an actual human subject, he persists in these (and certain other) misrepresentations of the adnexed vessels (Figs. 32 and 33).

A comparable but somewhat different error is made by the three anatomical illustrators who were contemporaries of Vesalius. Berengar in 1522-3 (Fig. 38), Estienne in 1539 or earlier (Fig. 12) and Dryander in 1541 (Fig. 39) agree in figuring the aortic arch with but a single branch, a *truncus brachiocephalicus*. This they present as the source of all four great vessels of the upper part of the body—that is, of both carotids and of both subclavians. Such was still the view of Fallopius when he issued his *Observationes anatomicae* at Venice in 1561.

These extraordinary errors of Vesalius (and his successors), and the comparable errors of his contemporaries, demand explanation. Before attempting to give it, we would point out that the anatomy of the great branches of the aorta are correctly figured by Eustachius. His investigations cannot be exactly dated, but were made about 1550, though not published till the eighteenth century (Figs. 19, 20).

The two divergent but erroneous descriptions of the great arteries by Vesalius on the one hand, and by his contemporaries Berengar, Dryander and Estienne on the other, are based on two divergent accounts given by Galen. These we may now consider.

(*a*) In his *De venarum arteriarumque dissectione* Galen writes :—" Thou seest that this artery [aorta], soon after leaving the heart, divides into two very unequal branches. The smaller divides immediately into two unequal parts, of which the greater [*truncus communis*] extends obliquely upward from left to right of the thorax toward the throat. The other [the left subclavian] goes in the opposite direction obliquely to the left scapula and axilla, sending branches to the sternum [internal mammary], to the upper ribs [superior intercostal], to the six vertebrae of the neck [vertebral] . . . and to the arm [brachial]. The great branch, passing upward toward the throat [*truncus communis*], reaches the [thymus] gland and produces a branch [left carotid] parallel to the left jugular vein, and after this another branch [right carotid] which accompanies the right jugular. All the rest of this artery [right subclavian] branches in the way which I have described for the branch to the left shoulder and axilla." *

* Chapter IX, Kühn, p. 818.

On this passage we have four remarks to make. Firstly, the arrangement of vessels described here is that of an ape (Figs. 34 and 44), though not far from that of a cat. In the ape the conditions differ from those found in domestic ungulates and

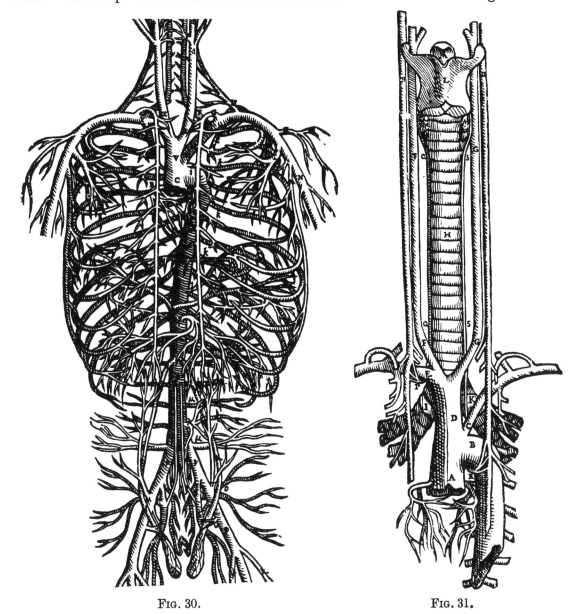

FIG. 30.   FIG. 31.

FIG. 30.—General arterial scheme from *Fabrica* p. 295 *. From aortic arch two branches arise close together: V, truncus communis, and D, left subclavian, as in the ape. The form of the fork between the innominate and left carotid, the large lateral thoracic arising near N, and the higher level of the right renal are all ape-like. The conspicuous parallel umbilical branches of the internal iliacs are foetal. The large diagonal artery at the side of the neck is not human. It represents the inferior cervical artery, which is particularly developed in the pig.

FIG. 31.—Branches of aorta, etc., from *Fabrica*, p. 328. The great vessels are ape-like. The relations of the recurrent laryngeal nerves to the vessels are well shown.

in the dog and, to a less extent, from those found in man. Secondly, this work of Galen on the dissection of veins and arteries was unknown in the West during the Middle Ages. No early MS of it has survived. Thirdly, it first appeared as a Latin

translation by Antonio Fortolo Joseriensis (? of Jesi), printed in France in 1528.*
Guenther's teaching on the veins and arteries was taken from it. Fourthly,
Fortolo's text was revised by Vesalius himself for the collected edition of Galen's
works known as the 'First' (really Second) Giunta edition, Venice, 1541-2.
It is the scheme which Vesalius follows in *Tabulae* and *Fabrica*.

(*b*) In his *De usu partium* Galen writes:—" This artery [the aorta] having
arisen from the left cavity of the heart, sends branches to all parts of the body of
the animal (*zoon*) and divides into two unequal branches. Of these, one is much
the greater [*aorta descendens*], for it is directed to the lower parts. Now the
parts situated below the heart in every animal (*zōon*) are much more numerous
and much greater than those above." †

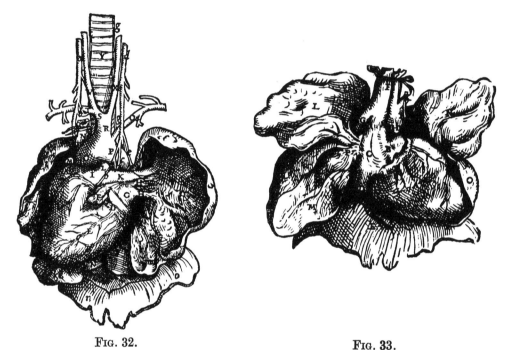

FIG. 32.   FIG. 33.

FIG. 32.—Thoracic organs with heart pulled to right, *Fabrica*, p. 564. A, B, C, left side of heart ;
  D, E, coronary vessels ; F, left auricle ; G, H, ' arteria venalis '= pulmonary vein ; I, K, ' vena
  arterialis '= pulmonary artery ; L, ductus arteriosus ; M, right auricle ; NN, vena cava ; O, P,
  aorta ; Q, ' Portion of arteria magna ascending to left axilla, and R, portion ascending throat ' ;
  S, ' Right part of the portion of the arteria magna ascending to throat, which gives off T,
  the axillary artery of the right arm ' ; V, X, ' Soporal or somniferous arteries '= carotids ;
  Y, ' arteria aspera '= trachea.

FIG. 33—Thoracic organs with heart pulled strongly to left, exposing vena cava. From *Fabrica*,
  p. 563. A, right ventricle ; C, vena cava, represented as a single vessel penetrating the
  diaphragm E at D, giving off the right atrium at B and the azygos at G, and passing up into
  the neck at F ; L, M, N, I are lobes of the lung ; H, the aorta.

On this passage we have again four remarks to make. Firstly, Galen is here
frankly and specifically describing the anatomy of an animal. The term (*zōon* =
animal) occurs twice in it. Secondly, the branch of the aorta that does not descend
is a *truncus branchiocephalicus,* as in domestic ungulates (and the dog), but not a
truncus communis, as in the ape (Figs. 34). Thirdly, the *De usu partium* was
widely read in the Middle Ages in the Latin translation by Nicolas of Reggio

* A few notes on Fortolo, of whom almost nothing is known, will be found in Cushing's
Biobibliography.

† Book VI, Chapter 5 ; Kühn, III, p. 428.

of about 1310. Some forty MSS survive. It was in general use in the medical schools in the sixteenth century, was well known in Paris, Padua and Bologna, and was edited in 1538 by Sylvius, who used it in his lectures which Vesalius attended. Fourthly, the account by the 'Arabs', Haly Abbas and Avicenna, is identical with that of Galen given here.* The European contemporaries of Vesalius also set forth this view of Galen. Vesalius, on the other hand, whenever he deals with the aorta, sets forth the views expressed by Galen in the *De venarum arteriarumque dissectione*.

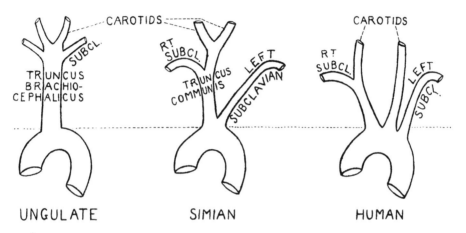

FIG. 34—Types of branching of aortic arch.

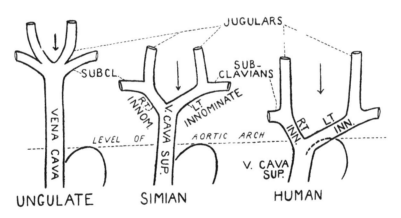

FIG. 35—Types of superior vena cava.

But was Vesalius merely following the account of the great arteries in Galen's *De venarum arteriarumque dissectione*? The answer must be, No. The figures of these vessels in the *Tabulae* and the *Fabrica* are not based solely on a text. Many of the drawings of the aortic branches in the works of Vesalius carry the conviction of being based on dissections. Even were Vesalius not known to have dissected apes, we should be confident that the draughtsman of these figures (or his guide) had access to actual anatomical structures. In the *Tabulae*, as in the *Fabrica* Vesalius is not merely visualising printed accounts : he has in his mind's eye what he had actually seen in apes.

* P. De Koning, *Trois traités d'anatomie arabes*, Leyden, 1903, p. 606.

The Vesalian errors concerning the aortic branches were remarkably persistent in the anatomical tradition. In the seventeenth century his figures of the branches of the aortic arch were still circulating. They appear in Caspar Bartholin's *Institutiones anatomicae*, edited by his son Thomas in 1647. This was the current student's text-book of the time.

§ 5. *Superior Vena Cava.*

In man the superior vena cava is formed at or just below the level of the top of the aortic arch, by the junction of two innominate veins. The left innominate is much more oblique than the right and almost three times as long (Figs. 35 and 45).

In the ape the superior vena cava is relatively much longer than in man. It is formed by the junction of the two innominates at a level well above the top of the aortic arch. This junction, as well as the main course of the vena cava superior. is nearer the middle line than in man (**Figs. 35 and 44**). Thus the two

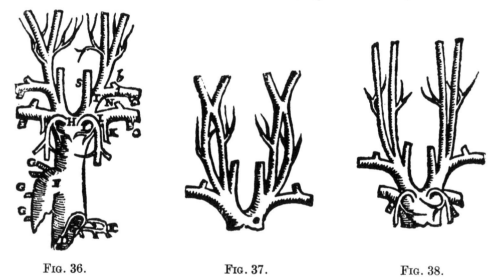

FIG. 36.   FIG. 37.   FIG. 38.

FIGS. 36–38.—Variations in tributaries of the vena cave superior, from *Fabrica*, p. 284.
Fig. 36 is of the ungulate type, Fig. 37 is near the ape type, Fig. 38 is
intermediate between the other two. None is human.

innominate veins in the monkey are almost equal and symmetrical. (In the cat the great veins are somewhat similarly arranged to those of the ape, but the innominates are longer and the vena cava superior correspondingly shorter.)

In the domestic ungulates (and in the dog) the innominates are hardly represented at all (Figs. 15 and 35). In them the great vein from the neck and that from the anterior limbs conjoin to form a vena cava at a level far above the aortic arch and almost symmetrically on the two sides.

In Tabula II Vesalius presents the ungulate type of superior vena cava. To this he adheres in the Venesection epistle of 1539 (Fig. 48). In the *Fabrica* and *Epitome* he hesitates between the ungulate type and something near to the simian type, favouring the ungulate (Figs. 36, 37, 38, 41 and 49).

Of the contemporaries of Vesalius, Berengar describes but does not figure the ungulate type (1522); Dryander figures but does not describe it (Fig. 40);

Estienne describes and figures it (before 1539, Fig. 9).   Fallopius describes it and figures it on the only occasion on which he invokes the graphic method (1550).

Again we can refer to the source from which all these writers draw.   Galen in his *De venarum et arteriarum dissectione* describes a great vertical superior vena cava into which all four great vessels open or, as he puts it, 'from which they are

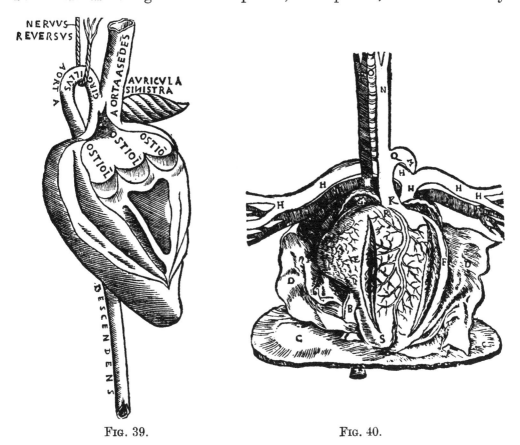

Fig. 39.   Fig. 40.

FIG. 39.—Heart from Berengar's *Isagoge*, Bologna, 1523.   The opened left ventricle shows semilunar valves much out of drawing.   The 'ascending aorta' is of ungulate type.   The auricle is 'cochlear' ('The cockles of the heart').

FIG. 40.—Heart, from Dryander, 1541, showing QN, truncus brachiocephalicus of ungulate type arising from aortic arch, and P, vena cava superior, also of ungulate type.   The legends read: "C, C, diaphragm; D, B, D, involucrum of heart, a thin membrane surrounding the heart and filled by its substance; E, right ventricle; F, left ventricle; G, auricles; H H H H and I I I, courses of pulmonary vein and artery [entirely misdrawn]; K, beginning of aorta; Q, N, aorta ascendens [*truncus brachiocephalicus*]; Q, M, aorta descendens; O, trachea; P, great ascending vein; R, veins nourishing heart; S, apex of heart."

given off '.*   The passage, though inaccessible to Berengar, was available to him indirectly in the *Canon* of Avicenna.†

Thus for his account of the superior vena cava Vesalius drew on the texts of Galen and on his own knowledge of animal anatomy in a somewhat similar way

* Chapter 2, Kühn, II, p. 787.

† P. De Koning, *Trois traités d'anatomie arabes*, Leyden, 1903, p. 622.

as for the great arteries (pp. xlvi–l). Basically his descriptions and figures of the superior vena cava are those of Galen and, therefore, of an ungulate. Vesalius has to some extent corrected this by reference to the bodies of apes (and cats), but hardly to the human subject. For the great arteries Vesalius thus ultimately depends on the ape, for the great veins on the sheep.

The Vesalian figures and description of the tributaries of the superior vena cava were even more persistent than this account of the aortic branches. We find it still in the last edition of the *Institutiones anatomicae* (Amsterdam, 1686) of Thomas Bartholin (1616–80). All anatomical works to the middle of the seventeenth century figure the trunk of the superior vena cava as symmetrical with symmetrical branches (tributaries).

From the comparative standpoint it has to be remembered that the symmetrical or medial position of the superior vena cava is a secondary feature, since the vessel is formed by the fusion of the two innominates which are of different origin. The right innominate is mainly a survival of the right anterior cardinal vein while the left is mainly a survival of a secondary communication between the two anterior cardinals.

§ 6. *Caval System.*

In Tabula II (as throughout the *Fabrica* and *Epitome*) Vesalius treats the two venae cavae, superior and inferior, as forming one continuous vessel. This great trunk runs near the mid-line of the body cavity through most of its length (Fig. 41). It swerves to the right at its source from the liver and to the left in the region of the heart. At the limit of the swerve to the left the right cavity of the heart comes off as a large blind branch of the cava itself. The main longitudinal trunk is treated as arising from the liver (*Jecur venarum principium*). It is presented as bifurcating above and below in a substantially similar manner. The azygos vein, however, is always enormously exaggerated by Vesalius, who developed peculiar views on its significance.

All this, though very far from the modern presentation, is not peculiar to Vesalius (except for the exaggerated azygos). The early anatomical writers are at one in their emphasis on the continuity of the inferior and superior venae cavae. All treat the two as one continuous vessel, from which the cavity of the right heart is offset. This represents a definite point of view which demands explanation.

In the adult human heart the entry of the superior vena cava into the right atrium is separate from the entry of the inferior cava. At their nearest approach the terminations of these vessels are about an inch apart, and between them extend the two right pulmonary veins, united together and to the ends of the venae cavae by a fold of pericardium. The ends of the venae cavae are connected externally by a shallow groove in the wall of the atrium, the *sulcus terminalis*. This marks externally the limits between the auricular appendage and the smooth-walled remainder of the right atrium.

The ape's heart presents certain differences from that of man (as well as from that of all the domesticated animals). In the ape the terminal portions of the superior and inferior vena cava appear, on external view, to be continuous with each other, being united by a *sinus intermedius* (Fig. 44, cp. Figs. 45 and 47) This appearance of continuity is associated with the fact that the right pulmonary veins are less intimately applied to the heart in the ape

than in man.   Another way of saying the same thing is that the pericardial fold
that links together the superior vena cava, the right pulmonary veins and the
inferior vena cava is further to the left in the ape than in man.

There was another anatomical source available to Vesalius, and doubtless to
Galen, to which reference is seldom made.   There must always have been difficulty

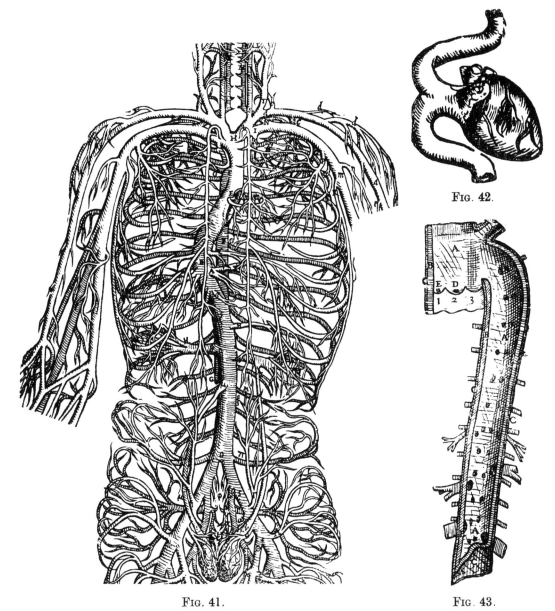

FIG. 41.

FIG. 42.

FIG. 43.

FIG. 41.—General scheme of venous system, *Fabrica*, p. 268*.   Compare with Tabula II.

FIG. 42.—Heart attached to the vena cava, which is severed above and below *Fabrica*, p. 275*,
Vesalius bids the reader to compare with Fig. 41, from which the heart has been removed
but the vena cava left intact.

FIG. 43—Aorta laid open.   From *Fabrica* p. 259*.   It is ape-like.   A A A, cavity of vessel ; B, the
inner and C the outer coat ; 1, 2, 3 mark the semilunar valves ; D and E are the orifices
from the two anterior aortic sinuses into the coronary arteries.

in disposing of the bodies of stillborn children.   These furnish handy anatomic
material and Massa says that he himself used them for the purpose (1536,
p: xxxiv).   Now the superior and inferior venae cavae of the full-time foetus
(Figs. 45 and 46), and even more of the earlier foetus (Fig. 47), give the impression

of forming one continuous vessel. The connexion of the cavae with each other is less obscured than in the adult by the pulmonary veins. The relations of the venae cavae of a foetus approach those erroneously figured and described by Vesalius as present in the adult (Tabula II and Figs. 33, 41 and 42).

The continuous vena cava, of which the right side of the heart forms a branch, was an essential element of Galen's physiology (see p. xxxix). It is for that reason

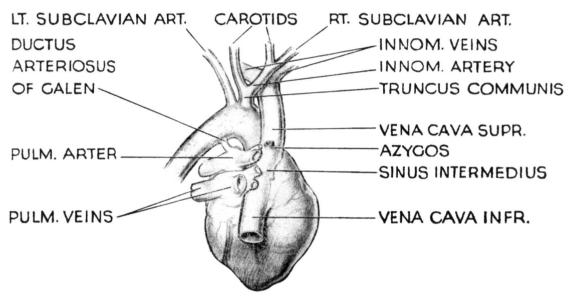

FIG. 44.—Heart and great vessels of young adult rhesus from behind. A *sinus intermedius* between the two venae cavae is conspicuous. Junction of innominate veins well above level of aortic arch. Short but definite truncus communis. Natural size.

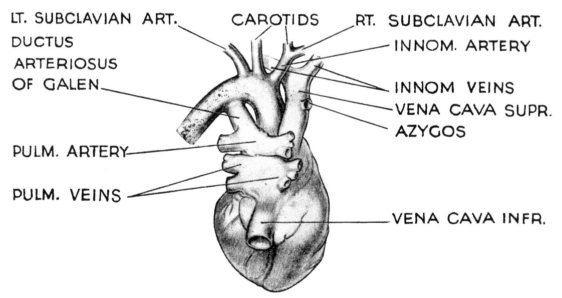

FIG. 45.—Heart and great vessels of full-term human foetus from behind. Continuity of the venae cavae obscured by pericardial fold. Junction of innominate veins at or little above level of aortic arch. No truncus communis. Natural size.

that all the sixteenth and most of the seventeenth century anatomists present this same feature. Estienne (Fig. 9) shows it well, as does Fallopius on the only occasion when he uses a diagram. Even Eustachius distorts the facts to represent the same feature (Fig. 20).

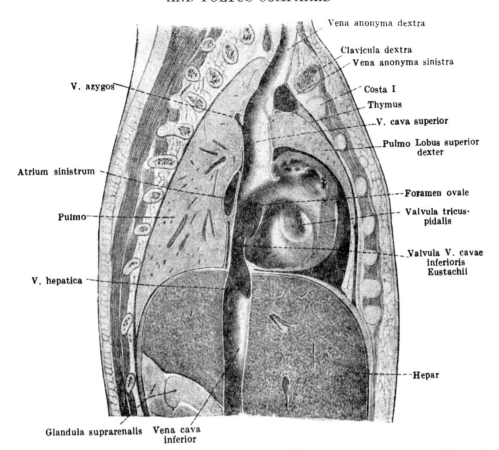

Vena anonyma dextra
Clavicula dextra
Vena anonyma sinistra
Costa I
Thymus
V. cava superior
Pulmo Lobus superior dexter
Foramen ovale
Valvula tricuspidalis
Valvula V. cavae inferioris Eustachii
Hepar

V. azygos
Atrium sinistrum
Pulmo
V. hepatica
Glandula suprarenalis
Vena cava inferior

FIG. 46.—Sagittal section of thorax of still-born child in plane of venae cavae, viewed from the right. 5/4 natural size. It shows the two venae cavae in line with each other. Eustachian and tricuspid valves and foramen ovale are visible. Right innominate vein in the same plane as the venae cavae. From J. Tandler, *Lehrbuch der systematischen Anatomie*, Band III, *Das Gefass-system*, Leipzig, 1926. The structure marked *Atrium sinistrum* is probably the common base of the right pulmonary veins.

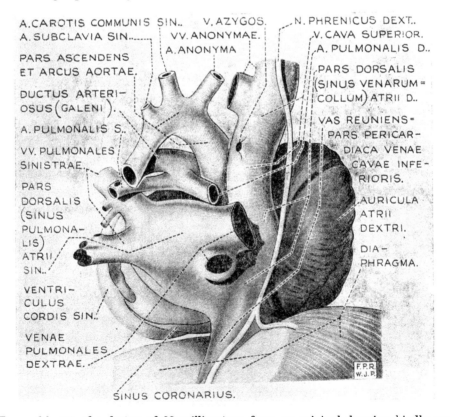

A.CAROTIS COMMUNIS SIN.
V.AZYGOS.
N. PHRENICUS DEXT.
A. SUBCLAVIA SIN.
VV. ANONYMAE.
V. CAVA SUPERIOR.
A. ANONYMA
A. PULMONALIS D.
PARS ASCENDENS ET ARCUS AORTAE.
PARS DORSALIS (SINUS VENARUM = COLLUM) ATRII D.
DUCTUS ARTERIOSUS (GALENI).
VAS REUNIENS = PARS PERICARDIACA VENAE CAVAE INFERIORIS.
A.PULMONALIS S.
VV. PULMONALES SINISTRAE.
AURICULA ATRII DEXTRI.
PARS DORSALIS (SINUS PULMONALIS) ATRII SIN.
DIAPHRAGMA.
VENTRICULUS CORDIS SIN.
VENAE PULMONALES DEXTRAE.
SINUS CORONARIUS.

FIG. 47.—Base of heart of a foetus of 93 millimetres, from an original drawing kindly supplied by Dr. Franklin P. Reagan. A sinus venarum of the right atrium (*collum atrii dextri*) intervenes between the two venae cavae. This is less obscured by the pulmonary veins than in the later stages of development, and recalls the situation in the ape (Fig. 44). The line of the right phrenic nerve here marks almost exactly the ' sulcus terminalis ', or line of separation, of the sinus venarum and auricular appendage. The dark line on the atrium to the right of this is the shadow thrown by the phrenic nerve.

This ancient view of the continuous vena cava with the right heart as offset was so firmly embedded in anatomical tradition that it long survived Harvey's discovery, with which it is inconsistent.   Thus Johan Vesling, professor of anatomy at Padua, still maintained it in his *Syntagma anatomicum* (Padua, 1647). It appears in the English translation of that work printed in London in 1651.   It may still be seen in an edition of Carlo Ruini's beautiful *Anatomia del Cavallo* as late as 1707 (Fig. 15).

§ 7. *Posterior Intercostal Vessels.*

The posterior intercostal veins as figured in Tabula II present several non-human features.   In man the first, second, usually the third, and rarely the

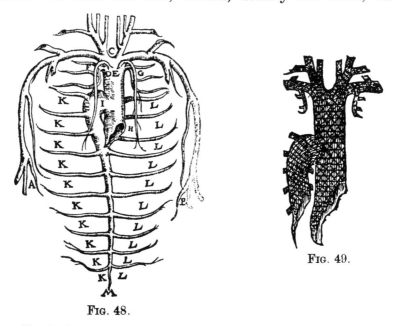

Fig. 48.

Fig. 49.

Fig. 48.—From Vesalius's *Venesection Epistle*, Venice, 1539.   Superior vena cava of ungulate type.   Three higher intercostals are represented as ascending as in the ape instead of two as in man.   "A, B, axillary veins cut off at the forearms ; C, the great bifurcation in the [thymus] gland or throat ; D, E are the veins running behind the *os pectoris* to supply the anterior thoracic wall [internal mammaries] ; F, G, the three veins nourishing the upper ribs ; H, the trunk of the vena cava cut off where it adjoins (*tangit*) the right auricle and gives off the right cavity of the heart ; I, unpaired vein [*azygos*] ; K, K, etc., are nine branches of the vein nourishing the lower nine ribs ; L, L, their companions nourishing the nine left ribs ; M, end of unpaired vein."   The azygos is greatly exaggerated and enters the vena cava on its right, in accord with the special view of Vesalius.

Fig. 49.—"The part of the vena cava between the right cavity of the heart and the throat, which we figure to display our doctrine of the fibres in the walls of the veins ", *Fabrica*, p. 259*. Vesalius here described these ' fibres ' as extending in four directions.

fourth, posterior intercostal veins unite to form a superior intercostal which on the right joins the azygos and on the left the innominate.   In Tabula II the upper two of the posterior intercostals join together on both sides to form a superior intercostal which on both sides empties into the vena cava itself.   This arrangement is not found in any domestic animal.   The figure errs further in that no hemi-azygos is represented.   The lower posterior intercostal veins are represented as emptying symmetrically into the azygos.

These errors are carried over into Vesalius's venesection epistle of 1539 (Fig. 48). They reappear in the *Fabrica* (Figs. 36, 38 and 49). In that work, however, Vesalius admits the occasional existence of a hemi-azygos, though he does not represent it in his general scheme of the veins (Fig. 41), in which he reduces the intercostal tributaries of the superior intercostal to two on each side.

In his account of the upper intercostal veins Vesalius draws from Galen's *De venarum arteriarumque dissectione*.* Galen there says specifically that he is describing the situation in the ape. Thus here again Vesalius had used simian anatomical features.

As regards the corresponding arteries, in man the subclavian artery gives rise to a costo-cervical trunk from which arise the first two, or occasionally the first three, of the posterior intercostal arteries. Tabula III here fairly represents the human distribution. The same is true of the *Fabrica*, with a reduction of costal branches to two on each side. Vesalius may be here relying on Galen's *De usu partium* (Figs. 30 and 31).

§ 8. *Relative Levels of Kidneys and of Renal Vessels.*

In man the right kidney is lower than the left. This is said to be related to the great size of the right side of the human liver. The renal vessels in man are correspondingly connected with the aorta and the vena cava respectively at a somewhat lower level on the right side than on the left. In the domestic animals—cat, dog, horse, sheep, ox, as well as in the ape and rodents (though not in the pig)—this relation is reversed. In all these creatures the level of the left kidney and its vessels is considerably lower than the right. The first three Tabulae and all relevant figures in the *Fabrica* † and *Epitome* show a higher level for the right kidney and/or its vessels than for the left (Figs. 10, 30, 41, 55 and 56).

The lower level of the right kidney and its vessels is evident at any autopsy of a human body. To make the right higher, as he always does, Vesalius must have had his anatomical picture fixed from the bodies of animals before he attacked the anatomy of man. Estienne makes the same error (Figs. 9 and 11). He and Vesalius have accepted the statement of Galen that ' in all animals the right kidney is at higher level than the left ', ‡ which is itself a repetition of Aristotle §. Again that admirable observer Eustachius avoids the mistakes of his contemporaries (Fig. 19). Later anatomists, such as Casserius, also avoid the erroneous placing of the kidneys and adnexa.

§ 9. *Five Equal Lobes of Liver.*

In Tabulae I and II the liver, with five equal lobes arranged more or less symmetrically round a central dome, is quite unlike the human structure. Moreover, it is quite unlike the organ of any domestic animal. It approximates to the conditions in the freshly exposed liver of an ape (Fig. 53). The *Fabrica* and *Epitome*, however, have a number of representatives of the organ based on human material, avoiding the five equal lobes.

* Chapters II and V ; Kühn, II, pp. 786 and 787.
† There is a possible exception in the *Epitome*.
‡ *De anatomicis administrationibus*, VI, Chapter 3 ; Kühn, II, p. 579.
§ *De partibus animalium*, 671 *b*, 25.

A dome with five equal lobes is the traditional medieval form of the liver (Fig. 52). In anatomical drawings before the sixteenth century it is almost invariably so represented. The idea does not come direct from Galen, for he says that the number of lobes differs in different animals, and suggests that there are four in man. Nor does the idea of a five-lobed liver come from the Arabian interpreters of Galen such as Avicenna. Galen says that the lobes of the liver spread out 'like fingers of the hand'. It may be that this phrase suggested to medieval writers the number five.

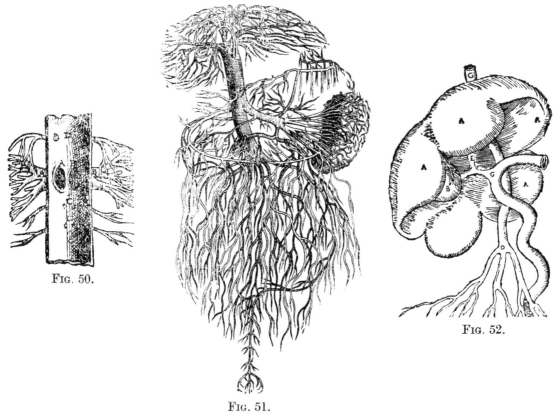

Fig. 50.   Fig. 52.

Fig. 51.

Fig. 50.—From *Fabrica*, p. 277. Inferior vena cava opened to show entry of hepatic veins. They are arranged in three groups—D, AB and C. In the Vesalian physiology these are the origin of the vena cava. The structure here figured is undoubtedly human.

Fig. 51.—Scheme of portal system, *Fabrica*, p. 262*. Venous tree in the liver obtained by maceration, as described by Massa (p. xxxv). Vesalius describes and figures the portal vein as dividing into five branches on entering the liver. This is a survival of the medieval convention of five lobes, see Fig. 52.

Fig. 52.—From Dryander, 1541. Liver with conventional five lobes. Gall bladder, B, is exaggerated in the medieval manner. D is the portal vein, C the intestine, G the inferior vena cava.

The liver in Tabula I has five almost equal lobes. It seems to be that of an ape drawn perhaps with the medieval diagram in mind. Some such sketch by Vesalius was doubtless copied by Dryander, who added errors of his own (Fig. 52). This may have been one of the clumsy plagarisms of which Vesalius complained.†

In the *Fabrica* Vesalius derides the idea that the liver has five lobes, pouring scorn on the lore and nomenclature that arises therefrom.‡ Nevertheless, he

† Harvey Cushing, *Bio-bibliography*, p. 16 ff.
‡ *Fabrica*, p. 506.

has not freed himself from these very things.   He says that the portal vein on arriving at the gate of the liver divides usually into five branches, which he figures and numbers (Fig. 51).   This is wrong for man but right for the ape. The facts were accessible to Vesalius in the work of a contemporary, for Massa (1536) had described a maceration technique for demonstrating the branches of the portal and the tributaries of the hepatic veins.   In fact the portal vein on entering the fissure of the liver divides into only two main branches (Fig. 59). This is presented correctly for the first time by Vesling in 1647.*   The portrayal of the portal vessels has hardly been improved since that of Francis Glisson in the first original work of anatomical importance printed in England (Fig. 58).†

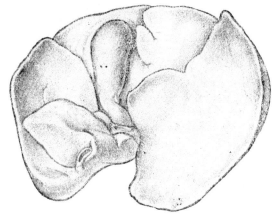

Fig. 53.—Lower surface of liver of young adult rhesus monkey, natural size.
Five lobes visible and substantially equal in appearance ; no preponderance of right lobe.

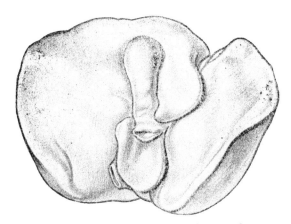

Fig. 54.—Lower surface of liver of full-term human foetus, natural size.
Large right lobe.

Galenic physiology lays great emphasis on the gall-bladder as the seat of one of the four humours, namely, the choler or yellow bile.   The bladder is thus always shown in medieval drawings as preternaturally large, as in Tabula I.

The cystic vessels on the surface of the gall-bladder in Tabula I show that in 1538 Vesalius was familiar with that structure by actual dissection, though not necessarily in the human subject.   The junction of cystic duct with common hepatic duct to form the bile duct can be seen, but the bile duct passes behind

* *Syntagma anatomicum*, Padua, 1647.
† *Anatomia hepatis*, London, 1651.

i 2

instead of in front of the portal vein. This is a serious anatomical error, corrrected in the *Fabrica*. It may be a mistake of the block-maker and not of Vesalius, who certainly made this drawing himself.

§ 10. *Homologies of Generative Organs of the Two Sexes.*

In the figures of the generative organs in Tabula I Vesalius seeks to illustrate homologies between the two sexes. To these he mainly adheres in the *Fabrica*. They may be tabulated thus :

FEMALE.	MALE.
Vulva.	Urinary meatus.
Vagina.	Penis.
(Collum uteri=cervix).	Neck of bladder.
Cornua uteri (round ligament ?).	Prostate.
Uterus.	Bladder !
Fallopian tubes.	Vasa deferentia.
Pampinniform plexus of ovarian vein.	Pampinniform plexus of spermatic cord.
Ovary.	Testis.

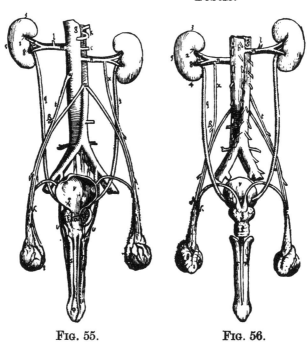

FIG. 55.   FIG. 56.

From *Fabrica*, p. 375. Male urinogenital organs.

Fig. 55 from in front, with the anterior wall of the bladder removed ; fig. 56 from behind. The right kidneys and renal vessels are higher than the left, as always in Vesalius. The prostate is properly figured for the first time—compare Tabula I. Ampullae of vasa deferentia are shown but not seminal vesicles. The sphincter is exaggerated. Five pairs of infra-renal lumbar veins are shown, but there are only three in man. Some ungulates have five.

Most of the erroneous parallelisms represent a tradition dating back to medieval times and ultimately to Galen.* The conception of *cornua* or 'horns' of the

* This tradition is discussed by F. Weindler, *Geschichte des gynäkologisch-anatomischen Abbildungen*, Dresden, 1908, and more recently by Charles Singer in *The Fasciculo di Medicina*, Venice, 1493 ; Florence, 1925.

uterus was peculiarly persistent. Even Leonardo figures them very clearly. It is difficult to connect them with any definite structure, but perhaps the round ligament with misunderstood connections may be hazarded. The parallelism of the bladder in the male with the uterus in the female is a specially gross error. It is peculiar to this diagram and to those that derive from it, including the monstrous figures of the female parts in the *Fabrica*.

§ 11. *Prostate Gland.*

Tabula I contains the earliest representation of the prostate. It shows two lobes in front. The organ had been first described in modern times in 1536 by Niccolo Massa but without figures. His book was printed at Venice and Vesalius can hardly have missed it.

In the small figure of the insertion of the vasa deferentia, these vessels are represented as passing behind instead of in front of the ureters; this is, presumably, an error of the woodcutter. The same figure shows an enlargement at the termination of the vas. This is doubtless the ampulla of the vas united with the seminal vesicle. More laterally the figure exhibits two lobes at the neck of the bladder, which must be an attempt to render the prostate gland. That organ is better represented in the *Fabrica* (Figs. 55 and 56), where it is described as ' a glandulous body surrounding the insertion of the vessels carrying down (*deferentia*) the semen '.

Vesalius in the *Tabulae* gives the organ no name. In the *Fabrica* he calls it alternatively *corpus glandulosum* and *assistens glandulosus*, translating Galen's ' **adenōdes sōma** '* and ' **parastatēs adenōdes** '.† It was first called ' prostate ' by Caspar Bartholin in 1611. † He applies the word wrongly, for Galen, relying on Herophilus, applies it to the epididymis. §

Tabula I shows that Vesalius had seen the prostate in the human body. His figure could not have been derived from the ox, sheep or carnivor, for all these animals have a small circular prostate. Nor could it have been drawn from the ape in which the prostate, though well developed, does not extend in front of the urethra.

§12. *Asymmetry of Spermatic Veins.*

In both figures of the generative organs in Tabula I the inferior vena cava is represented as greatly expanded, so that it hides most of the aorta. It is so shown throughout *Fabrica* and *Epitome* (Figs. 10, 55 and 56). The course of veins from the gonads themselves is correctly depicted for both sexes, that of the right entering the trunk of the vena cava and that of the left the renal vein. This observation has a peculiar history.

In 1536 Guenther published his *Institutiones*. In it he says : " The seminal veins differ from one another in their origin [that is what we call their termination]. The vein on the right takes its origin not from the side of the cava itself, but almost

* *De usu partium*, XIV, Ch. 9 ; Kühn, IV, p. 182.
† *Definitiones medicae*, LIX ; Kühn, XIX, p. 362.
‡ *Institutiones anatomicae*, Wittenberg, 1611.
§ *De usu partium*, X, Ch. II ; Kühn, IV, p. 190.

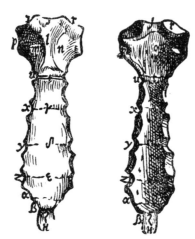

Fig. 57.—Sternum, front and back, from *Fabrica*, p. 87.   In man the sternum is formed of six segments, united in the adult, in apes of seven.   Vesalius here represents seven, but in the text (p. 89) implies that he has not seen it of more than six.   In Tabula IV it is clearly of seven, and is said to be *lunatum*—that is, with curved sides, the curves being very marked.   They are still over-emphasized in the *Fabrica*.

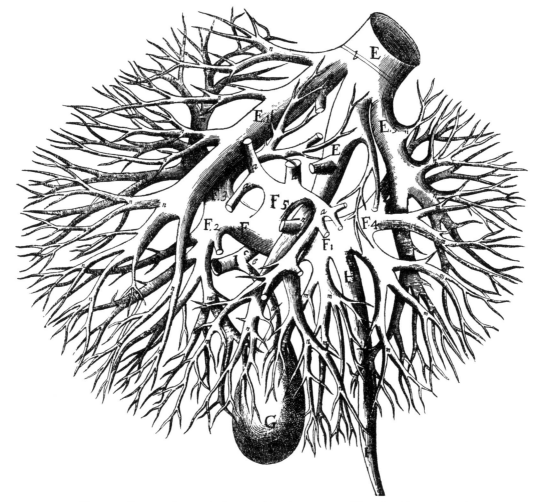

Fig. 58.—From Francis Glisson, *Anatomia hepatis*, London, 1654. "Vessels of liver from front. E, trunk of vena cava above diaphragm ; F, sinus portae ; F_1–F_5, ' branches ' of portal vein ; G, gall bladder ; H, H, umbilical vein ; I, common bile-duct, a, a, a, minor branches of F_5 ; b, part of diaphragm attached to cava ; c, bile-duct ; d, cystic duct ; e, junction of bile- and cystic-duct ; m, m, m, minor branches of porta ; n, n, n, minor branches of cava."

from the middle, below the [vein] which goes from there to the kidneys. That [vein] which reaches the testicle on the left side emerges from that [vein] which goes to [that is comes from] the kidneys. This I have not found previously recorded (*scriptum*) by any anatomists, and I believe not even noticed. But recently we have hit upon them through the skill of Andreas Wesalius . . . . Anatomists have neglected to trace out their origin, being satisfied that few if any veins go to the lower parts unaccompanied by an artery."

It is remarkable that Guenther should ascribe to Vesalius the discovery of an anatomical relationship that was in fact a medieval commonplace. In 1536, the very year of publication of Guenther's *Institutiones*, Massa (c. 1480–1569) issued his *Anatomiae liber introductorius*. It devotes four and a half pages to a correct account of the distribution of the seminal vessels. These had already been well described by Berengar in 1522 and by Mundinus in 1316. Mundinus

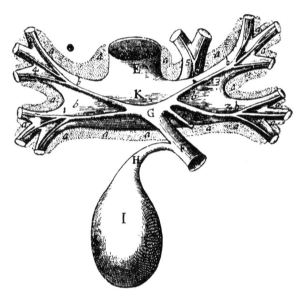

Fig: 59.—From F. Glisson, *Anatomia hepatis*, London 1654. " Main branches of portal vien and biliary passages from behind. *E*, trunk of porta pulled back ; 1–5, five major branches of porta ; *F*, common bile-duct ; *G*, point of primary division of common bile-duct ; *H*, cystiic-duct ; *I*, gall-bladder, *a, a, a*, common capsule [of Glisson] opened ; *b, b, b*, subdivisions of bile-duct."

refers Guenther's very point to Avicenna. Doubtless Avicenna himself derived it from a passage in Galen's *De ven. art. diss.* (Ch. VIII ; Kühn, II, 808), which Guenther missed.

Why, then, did Guenther give this undeserved credit to his pupil ? The answer must be ignorance of some and forgetfulness of other of these sources. Paris of his day was out of touch with the Italian medical schools. Guenther spent most of his time in translating medical works from the Greeks. He knew the Arabian writers well but had missed or forgotten the Avicenna references. Mundinus was never read at Paris and Guenther may not have seen his text.

Even accepting this ignorance, the introduction here by Guenther of the name of his prosector seems a little forced. It is perhaps an attempt to press the claims of a favourite pupil for some appointment—a ' testimonial ' in fact.

But apart from all this, an interesting connection here arises. Avicenna's knowledge of the asymmetry of the renal veins can be traced to Book XIII of Galen's *De anatomicis administrationibus*. Now this thirteenth book was unknown to Guenther though he had printed a Latin translation of the first nine at Paris in 1531. The last seven books were early lost from the Greek text. They survived only in an Arabic translation and were recovered in the twentieth century.* Thus on this particular point the sources of Mundinus and the medieval Arabists were actually better than those of the renaissance scholars, for they went back through Avicenna to elements in Galen that were otherwise inaccessible.†

§ 13. *Seven Segments of Sternum.*

In Tabula IV the sternum is badly drawn and its lower part is abnormally expanded. This structure is represented as divided into seven segments by six sutures and in the accompanying text Vesalius draws attention to this. But the sternum is in fact ossified from six (sometimes five) centres or pairs of centres, never from seven, either in man or the anthropoids. In *Macaca*, however, it arises from seven centres or pairs of centres. We have here a memory of simian anatomy which is carried over into the *Fabrica* (Fig. 57) derived in the first instance from Galen's *De ossibus*, ‡ though Vesalius is obviously drawing a human sternum.

---

* Max Simon, *Sieben Bucher Anatomie des Galen*, 2 vols., Leipzig, 1906 (see vol. I, pp. 135–7).

† Charles Singer, *Bulletin of the History of Medicine*, Baltimore, 1925, XVII, p. 426.

‡ *De ossibus ad tyrones*, Ch. XIII ; Kühn, II, p. 763 and elsewhere.

VI. Renaissance Anatomical Vocabulary.

§ 1. *General Sources of the Vocabulary.*

The language of the *Fabrica* is a highly involved, difficult and sophisticated Latin idiom of a type affected by the more pretentious mid-sixteenth century humanists. For scientific purposes it would be hard to find a more unsuitable medium. It is neither clear nor beautiful, neither expressive nor concise. It is claimed that the vernacular was still too unripe for scientific use. It may be rejoined that the idiom of Vesalius is over-ripe to rottenness. Happily the *Tabulae* confronts us with little of this irritating mechanism for the concealment of thought. The long dedicatory preface, which we have abbreviated to render it tolerable, illustrates the worst literary taste of the time. Its language is essentially a written, not a spoken, medium. Fluent though Vesalius may have been, he could not have spoken such tortuous, super-elaborated, repetitious rhodomontade.

It is, however, with the vocabulary that we are here concerned. The *Tabulae* is an almost ideal document for the study of renaissance anatomical nomenclature. To understand this technical terminology we must look into the literature from which it derived. The origins of the anatomical terms are in part medieval and in part 'modern'—that is, sixteenth century. The medieval elements are derived from two sources : (*a*) Classical, greatly corrupted and altered during the earlier medieval centuries and further changed by misunderstandings during the later medieval centuries ; and (*b*) Arabic with strong Hebrew linguistic influence, dating from the eleventh century onward and reaching the West through translations of Arabic medical works. The 'modern', *i.e.*, sixteenth-century, anatomical vocabulary has two sources comparable to those of the medieval : (*a*) Classical, through the Greek and Latin texts newly recovered during the revival of learning ; and (*b*) Arabic and Hebrew, derived by contemporary scholars, almost exclusively Jewish, from Arabic and Hebraeo-Arabic writings.

It is convenient to defer to a separate section (p. lxxix) the consideration of the Arabic and Hebrew elements, both of medieval and of 'modern' origin; and it is further convenient to consider in the present chapter the modern before the medieval sources of anatomical nomenclature. It will suffice for the moment to recall that Vesalius had access to a rich medieval anatomical vocabulary which he uses freely in the *Tabulae*. That work gives a very full list of anatomical synonyms. Their presence give it a special illustrative value in tracing the passage of the medieval into modern anatomical usage. The nature of the medieval anatomical vocabulary will emerge as the discussion proceeds.

§ 2. *Direct Classical Sources of the 'Modern' Vocabulary.*

We turn then first to the classical sources of the sixteenth century anatomical vocabulary. To understand the situation we must revise certain conventionally recognised historical sequences. The accepted account of the date and manner of recovery of the literary classics does not fit exactly to the record of recovery of the medical and anatomical works of antiquity. The conventional picture has to be modified in several particulars:

(*a*) The establishment of good texts was less strikingly epochal for medicine than for literature. A few Latin translations of medical works direct from the Greek were trickling into the West even before the thirteenth century, and

k

continuously in the thirteenth, fourteenth and fifteenth centuries. Moreover, there were differences in this matter between the Italian and the French schools, which had little inter-communication.

(*b*) For anatomy there was little 'revival of learning' in the fifteenth century. The anatomical texts were among the last important Greek texts to be recovered. The general anatomical reading of, say, 1520 was thus about as scholastic as that of 1320. It was perhaps even more so. In any event the study of the very few newly-recovered anatomical texts was a local affair that hardly affected most medical schools until about 1540.

(*c*) The familiar emphasis on the value of Greek studies tends to obscure the fact that the proportion of medical men who had facility in Greek, as distinct from a showman's acquaintance with it, was minute. The romantic recovery of Greek manuscripts was relatively unimportant for the advance of medicine. The pedestrian process of translation into the Latin, which all could read, was more directly significant. There is no adequate evidence that Vesalius himself was, in any effective or independent sense, a Greek scholar. Like the other practical anatomists of his century, he worked normally on Latin translations of Greek texts.

(*d*) As to date of first appearance in print, classical medical, and especially anatomic, texts were very belated as compared with the literary and philosophic. Galen was not available in Greek till 1525 and Hippocrates not till 1526. By then the corpus of Greek literature had long been available to scholars. This lag in the scientific texts is a factor in the divorce of the 'Humanities' from the 'Sciences'. Vesalius was in the happy position of being in the first generation to receive the full classical anatomic tradition.

(*e*) The mark of the spirit of science is neither recognition of good texts nor recovery of better. Even verification of the superior value of the better by appeal to Nature is but a preliminary. The only test that the scientific spirit is alive is the systematic and independent observation of the object—independent, that is, of any text, good or bad. We find hardly any evidence of a living scientific spirit among the anatomists until the sixteenth century is well on its way. The 'Humanists' were no more scientific than the 'Arabists'; perhaps they were even less so. Passing by the work of the artists and that of the inexplicable Leonardo, observational anatomy begins, in effect, with Berengar about 1520.

It is not hard to trace the influence of the impact of Greek medical learning on anatomy since few ancient texts and only four ancient authors seriously affected the anatomical tradition of the sixteenth century. These authors were Aristotle, Celsus, Galen and Pollux. The last takes a special place and needs separate consideration (see p. lxviii).

The passage of the Aristotelian corpus to the Latin West is an extremely intricate theme.* Happily we are not concerned with it as whole. It was

---

* The standard account of the Latin Aristotle is still that of Charles and Amable Jourdain, *Recherches critiques sur l'age et l'origine des traductions latines d'Aristote et sur des commentaires grecs ou arab es employés par les docteurs scolastiques*, second edition, Paris, 1843 ; M. Grabmann, *Forschungen über die Lateinischen Aristoteles-Uebersetzungen XIII Jahrhunderts*, Münster, 1916, supplements Jourdain. An important modern thesis is S. D. Wingate, *Medieval Latin Versions of the Aristotelian Scientific Corpus, with special reference to the Biological Works*, London, 1931.

only the three great biological treatises of Aristotle that affected the anatomical outlook. A conflated version by Michael the Scot from the Arabic of Aristotle's *Historia animalium, De partibus animalium* and *De generatione animalium* appeared very early in the thirteenth century. This remained current until after the middle of the sixteenth century. There is evidence that Vesalius had read it. Many commentaries on it passed into circulation. Of these the most popular were those of Albertus Magnus (1206–80), prepared about 1250, which Vesalius probably knew. Good versions from the Greek of all three biological works were made in or about 1260 by William of Moerbeke (1220–86). He had visited Constantinople at the instance of St. Thomas Aquinas (1227–74) and learned the language expressly for this task. William's version, however, by no means displaced Michael's, which long remained the favourite.

In the fifteenth century many scholars rendered Aristotle into Latin. Several Italianate Greeks applied themselves to the biological works. Chief of these was Theodore Gaza (1400–76) of Salonica. He came to Italy in 1430, a fugitive from the advancing Turks, and, after professing Greek at Ferrara, was brought in 1450 by Nicholas IV (Pope 1447–55) to Rome. There he made an admirable version of Aristotle's zoological works. It was magnificently printed at Venice in 1476 and often subsequently. This standard and scholarly version could not fail to influence Vesalius as it did the other anatomists of the sixteenth century. Traces of its vocabulary can be demonstrated in the *Tabulae*.

The most significant medical text recovered in the fifteenth century was that good common-sense summary of practice that bears the name of Celsus. Its last two books form a compendious surgery and include a fair account of the bones. MSS of it had been known since 1426, and had roused the interest of the great Politian (1454–94). When the beautifully printed text of Celsus appeared at Florence in 1478 it caused considerable sensation, for it was the first ancient medical work to be thus presented. Many editions followed which exerted much influence on anatomical nomenclature. Vesalius had read Celsus carefully. A number of the terms of the *Tabulae* are drawn from it.

Of works ascribed to Galen hardly any with anatomical bearing were printed in the fifteenth century. The only 'incunable' collection of the works of Galen was edited by Diomede Bonardo of Brescia at Venice in 1490. Of the hundred and fifty or so works to which the name of Galen has been attached this edition contains some seventy, but only three of them are anatomical, and two are spurious. (See § 3 under (*b*) and (*c*)).

§ 3. *Anatomical Incunabula.*

It will be useful to tabulate the printed anatomical texts available at the end of the fifteenth century. All are in Latin unless otherwise indicated :—

- (a) *Aristotle.* Biological works in 5 editions, all of Venice, beginning 1476 and including one in Greek of 1497. Also the *De anima* in 7 editions, beginning Padua, 1472.
- (b) *Galen, De anatomia matricis.* One edition by Bonardo, Venice, 1490.
- (c) *Pseudo-Galen, De juvamentis membrorum* and *De anatomia oculorum.* One edition by Bonardo, Venice, 1490.
- (d) *Celsus, De Medicina.* 4 editions beginning Florence, 1478.

(e) *Avicenna, Canon.*   15 editions beginning Padua (?) 1472 and including the Hebrew edition of Naples, 1491, used by the Jewish adviser of Vesalius for the *Fabrica.*   Also the *Cantica,* in 2 editions, Venice, 1483–4, which contains some vein-names.

(f) *William of Saliceto, Chirugia.*   7 editions beginning Venice, 1474.   In Latin (3 times), Italian (3 times), and French (once).

(g) *Guy de Chauliac, Chirugia.*   10 editions beginning Lyons, 1478.   In Latin (twice), Italian (twice), French (3 times) and Spanish (3 times).

(h) *Mundinus, Anothomia.*   9 editions beginning Pavia, 1476.   All in Latin except that in the great Italian *Fasciculo,* Venice, 1493.

(j) *Rhazes, Liber Almansoris.*   3 editions beginning Milan, 1481.

(k) *Haly Abbas, Liber regius.*   One edition, Venice, 1491, and *Liber continens,* one edition, Brescia, 1486.

The list contains the names of three ancients, three 'Arabs' and three medieval 'Arabists'.   There are 28 editions of medieval anatomists (including two pseudo-Galenic treatises).   Apart from Aristotle, the ancients are represented only by Celsus and by one edition of one minor work of Galen.   None deals with anatomy in the strict sense.   The only work in the list devoted entirely to anatomy proper is Mundinus.   There are 26 editions of Arabists, including Mundinus, and 20 editions of translations of Arabic writers.   There are 16 vernacular editions, all but one surgical.   Surgeons were often ignorant of Latin, and the great majority of medical works of the fifteenth century were for physicians and contain nothing that we should call anatomy.   For this there is good reason.   Until Vesalius had spoken, physicians thought in terms of humours and temperaments, always with the astrological scene in the background.   These concepts had very little to do with bodily structure.   It was the *Fabrica* which made medical men anatomically minded.

The absorption of the newly-won anatomical vocabulary of classical antiquity took place in a peculiar psychological atmosphere.   There was general revulsion against the 'barbarous' Latin of the Middle Ages, which had become almost a vernacular.   'Barbarous' terms were replaced, where possible, by their classical equivalents.   Sometimes, however, they were recast, often wrongly, into a classical form.   The texts were constantly 'castigated' and the vocabulary was continually 'purified'.

For the 'purification' of anatomical nomenclature and the substitution of Greek for Arabic and Hebraic terms a very important agent was a work to which even specialist scholars now seldom turn.   This is the *Onomasticon* of Julius Pollux (A.D. *c.* 134–92).   It was unknown in the Middle Ages, was first printed at Venice in 1502, and reprinted at Florence in 1520 and at Basel in 1536.*   The *Onomasticon*—the word means simply 'vocabulary'—was dedicated to the Emperor Commodus, son and heir of Marcus Aurelius and patient of Galen.   It consists of a series of lists of important words relating to various subjects.   Attached to these words are short explanations, often with quotations from ancient writers.   There is an elaborate and very long list of anatomical terms.   From this the humanist physicians often drew words to replace 'Arabist' terms in current use.

---

* There is a modern edition of the Greek text of Pollux by E. Bethe, Leipzig, 1900. We have found the edition of Wolfgang Seber, 2 vols., folio, Amsterdam, 1706, more useful for the study of the passage of the classical anatomical vocabulary.

Pollux became the source of many anatomical terms still in current use, for example *Amnion, Anthihelix, Antitragus, Atlas, Axis, Canthus, Clitoris, Cricoid, Epistropheus, Gastrocnemius, Tragus* and *Trochanter*.

§ 4. *Contributions to the Vocabulary from Renaissance Anatomies.*

A vernacular fifteenth century anatomy the vocabulary of which deserves study is that of Hieronymo Manfredi (1430–93) of Bologna. His work is of linguistic interest because it was composed for the entertainment of an educated man with no technical knowledge.* It therefore contains many current popular terms, a theme that has not been sufficiently explored. The anatomy of Manfredi's contemporary Giorgio Valla (1430–99), of Venice, on the other hand (p. xxix), is significant as the earliest work based on the newly recovered Greek vocabulary. Valla was familiar with the *Onomasticon* of Pollux. He must have studied it in MS since it was not printed until after his death. Among the terms in Valla's anatomy for the first time in Latin are : *Acetabulum* of the innominate, *Carpus* and *Metacarpus* replacing *rascetta* and its variants (notes 152, 178), *Cremastar, Diaphragm, Epigastrium, Epiglottis* (in the modern sense), *Hypochondrium, Hypogastrium, Mesentery, Olecranon, Omoplatum* for shoulder-blade (passing into French as ' omoplate '), *Poples* and *Popliteal, Psoas, Pylorus, Sura* and *Sural, Thenar* and *Hypothenar, Xiphoid* and *Zygoma*.

The *Anatomice* of Benedictus (1470–1525), which appeared at Venice in 1502, is less exclusively based on Greek sources than Valla's. It retains a considerable number of Latino-Arabic terms. On the other hand it is longer and more detailed than Valla's work and was much more widely read, especially in Paris. Among the terms that we trace in Benedictus for the first time, other than those he draws from Pollux, are *Chorioid, Axis vertebra, Bregma, Ethmoid, Glans penis, Hymen, Malleolus* for Malleus, *Peritoneum* and *Sesamoid*.

Of Gerbi and Achillini (p. xxx) we need say here only that their anatomical terminology is much as in Mundinus. We have traced no influence of either on Vesalius. It seems unlikely that the positive mind that constructed the *Fabrica* would endure the empty argumentation of their works.

Berengar (pp. xxxii-iii) was the most experienced of the early dissectors. His *Isagogae* of 1522/3 is one of the best sources for studying the modernisation of the anatomical nomenclature. As Berengar's anatomical methods stand at the turning point from medieval to modern, so his anatomical vocabulary marks the change from scholasticism to renaissance. Other writers before him, as Valla and Benedictus, draw their terms from the new-found Greek texts. The special feature of Berengar's *Isagogae* is that in it a practical man is using the new terminology not for learned display, but as a technical instrument to expound his own craft. This makes it easy to forgive an occasional solecism, for he is not preoccupied with correct classic models. Nevertheless Berengar was well read both in the Greek and Arabian literatures of anatomy.

About half of the vocabulary of Berengar is still medieval, drawn largely from Mundinus and from the translations of the Arabic writings, especially of Avicenna. Interesting and useful terms which he retains from the Arabs are, for example, *Alchatim*=hollow of sacrum (see note 332), *Bucella*=small bony eminence entering an articulation, *Nucha* = nape of neck and/or the associated spinal cord,

* The text of Manfredi's anatomy is printed in full and partly translated by Charles Singer, *Studies in the History and Method of Science*, vol. I, Oxford, 1917.

Nocra=hollow of neck, *Os cahab*=malleolus (see note 172), *Venae guidez*=jugulars (see note 40) and *Zirbus*=omentum.   Like all writers of the time, including Vesalius, he is very conscious of the nomenclatory confusion in his subject.   On account of this he usually gives several names for any structure.   Perhaps the confusion is greatest with the names of veins and especially of those commonly used for venesection.   The following passage, describing the veins of the forearm and hand, illustrates this.

" In these figures can be seen the course of the *salvatella* of Mundinus and Rhazes, the *sceylen* of Avicenna and the *salubris* of Haly and that branch of the *basilica* which ends between the little and ring finger which Rhazes calls [also] *salvatella*.   It can be seen moreover that the vein called *funis brachii* by Avicenna ends around the middle finger in a branch called *sceylen* by Haly and how also the *funis brachii* is a branch of the *cephalic* and terminates between the index and pollex."   This passage may be compared with the text of Tabula II under the letter V and the corresponding notes 51 and 54.

To give an idea of the vocabulary of Berengar we list from his index the technical terms entered under the letter A.   They give a fair sample of the whole :—

> *Abdomen*, classical and medieval.
>
> *Acrophallum*=**akromphalion**=middle of navel, probably from Benedictus.
>
> *Acrusta*=lumbar vertebra.   The word, which must be some compound of **akros,** is ascribed by Berengar to Homer !
>
> *Adjutorium*=*humerus*, a usual medieval term used by Mundinus.   See note 132.
>
> *Aegroides*, Greek, meaning ' like a goat-skin ', applied by Berengar to the tunica vaginalis of the testis.   We have not traced its origin.
>
> *Alaria*, *alulae*=*alae nasi*, medieval.
>
> *Alchael* and variants=hollow of the neck.   From the Latin Avicenna, Arabic *AL-KĀHIL*=interscapular region.
>
> *Alchatim*=hollow of sacrum.   See notes 321 and 332, Arabic *AL-QATN* =lumbar region.
>
> *Alsahic*=amigdalae.   Arabic *AL-FĀ'IQ*=hyoid bone.
>
> *Almochatim*=pericranium.   Arabic perhaps=*AL-KHĀTIM*=signet ring.
>
> *Amigdalae*=almonds=tonsils, a Greek word, but not in Galen or Pollux.
>
> *Anastomosis.*   Galenic.
>
> *Ancha.*   See note 239.
>
> *Animella*=thymus gland, an Italian and **Spanish** dialect word applied especially to calves.
>
> *Anteron* or *antheron*=chin, classical Greek.
>
> *Antiados*=tonsil, **antias, - ados** Galen, Hippocrates, etc.
>
> *Ante carpum*=*metacarpus*, medieval.
>
> *Aqualiculus*=lower abdominal cavity, medieval.
>
> *Arca aeris*=buccal pouch, recognised as prominent in apes ; **probably** Berengar's own term.

Arteriae sempiternae=persistent arteries=the umbilical arteries. A late medieval term.

Arteria aspera=*a. vocalis*=*a. trachea*. All forms are medieval, *aspera* being perhaps commonest, see p. xxxviii.

Arteria aorta. See note 71.

Arteria venalis. See note 85.

Arundines=tibia+fibula. See note 263.

Ascella=*ascellaris*. Medieval; first in Stephen's Haly Abbas (1127).

Auricula cordis, medieval and Celsus.

Auritium=semilunar valve, local Bolognese medical term = *aurithia* (see note 73).

Another sixteenth century Italian anatomist whose writing was known to Vesalius in 1538 is Massa (p. xxxiv). His work of 1536 is of great practical but of less linguistic significance. He was not very deeply read. His vocabulary is somewhat similar to that of Berengar but more definitely turned away from the Arabs and towards the Greeks. Words or definitions did not interest him, and he is of little importance for terminology.

The effective founders of anatomical nomenclature are Guenther and Sylvius, acting largely through Vesalius. Their teaching and their works were rightly the most popular of this time. Guenther has transmitted to modern anatomy a multitude of terms that he drew from his master Galen. Through him Galen is an influential anatomist at this very day. Among terms that Guenther seems to have introduced to the Latin anatomical vocabulary are *Coronary* process, *Dartos*, *Interseptum*, *Pericranium*, *Pinnula*, *Urachus* (replacing *Urinaculum* and its variants), while he popularised *Colon* as a substitute for *Longaon* and its variants.

The example of Sylvius is almost as living and active for modern anatomical teaching as is that of Vesalius for modern anatomical research. Sylvius was a master of method. We still use the instructional technique of which he was the first and one of the ablest exponents. Perhaps this comes out best in his systematic classification of muscles according to their attachments. His way of numbering blood vessels was no less admirably clear. Many names that he used for vessels have come down to our time. Among them are the *crural*, *cystic*, *gastric*, *intercostal*, *mammary*, *popliteal* and *sural* veins, and the *axillary*, *cervical*, *coeliac*, *iliac*, *mesenteric*, *pudenal* and *subclavian* arteries.

The modern anatomical vocabulary was formed in its early stages by the Parisian and Paduan schools. The *Tabulae* is, among other things, a dictionary of the new terminology introduced by these schools, confronted with the older anatomical terms. The antiquarian interest of Vesalius has preserved the latter for us. Beginning with the *Tabulae* and working through the later Paduan anatomies, it would be possible to present a coherent picture of the evolution of anatomical nomenclature. The *Fabrica* is the first great work of modern observational science, and some light is thrown on it by a terminological examination of its forerunner. The development of a suitable linguistic medium is a very important factor in the rise of science.

VII. Semitic Elements in the 'Tabulae'.

§ 1. *Introduction.*

In glancing at the *Tabulae* the eye is struck by a sprinkling of words in Hebrew characters. Most of these are Hebrew words but some are Arabic or arabicized words in Hebrew letters. Moreover the text displays throughout, printed in Latin letters, a number of words of Arabic origin and some of Hebrew origin. Such Semitic terms, spelt both in Hebrew and Latin letters, are more numerous in the last three tabulae, which depict the bones, than in the first three, which are concerned with the blood vessels.

The medieval Latin anatomical vocabulary contained, as is well known, many words of Semitic origin. As long ago as 1879 Joseph Hyrtl surveyed the subject in his *Das Arabische und Hebräische in der Anatomie.* Since then our knowledge has grown and Hyrtl's important pioneer work now greatly needs revision. In 1922 A. Fonahn of Oslo cast a yet wider net in his *Arabic and Latin Anatomical Terminology, chiefly from the Middle Ages.* This very useful list indicates the extent and difficulty of the field, but is merely a preliminary survey. Charles Singer broached the subject again in 1925 in examining the Semitic vocabulary of the leading medieval anatomist Mundinus.* Owing to war conditions the work of T. M. Peñuela did not reach us in time for use here.†

The Semitic influence on anatomy in general is especially seen in the medieval nomenclature of blood vessels and bones. These happen to be precisely the structures treated in the *Tabulae.* Before turning to these terms it will be well to review the course of Arabic and Hebrew linguistic influences on western anatomical nomenclature as a whole, with special reference to such Semitic anatomical terms as had reached Italy by the time of Vesalius.

§ 2. *History of Arabic Anatomical Terminology.*

In the seventh century the Arabs first entered into the heritage of the ancient civilizations of Byzantium and Persia. From their desert home they brought no scientific contributions. Moreover in the Byzantine and Persian Empires, where Islam spread rapidly, Greek science was at a low ebb save among the Syriac-speaking Nestorians. A flourishing medical school existed during the sixth and seventh centuries at the Nestorian metropolis, Gondisapur. Thence during the Umayyad period (661–749) learned men, and especially physicians, migrated to Damascus, the Islamic capital. They were mostly Nestorian Christians or recent converts to Islam.

The rise of the Abbasid Caliphs (750), with Bagdad as their capital, inaugurated the epoch of greatest Islamic splendour and prosperity. The thought of Islam was still, however, in the absorptive period. Medical literature in the Arabic language did not make its first appearance until the reign of the Caliph al-Mahdi (755–85), and then only with translations from the Greek. None of the early translators were Arabs and very few even Moslems. Most were Syriac-speaking Christians resident in Bagdad. A few were Jews. The Arabic versions were seldom made direct from the Greek and usually passed through Syriac.

* *The Fasciculo di Medicina, Venice, 1493, with introduction, etc., by Charles Singer,* two vols., Florence, 1925.

† *" Die Goldene " des Ibn al-Munāsif, ein Beitrag zur medizinisch-arabischen Lexikographie, etc.,* Rome (Institutum Biblicum), 1941.

Important work in preparing the ground for this transmission was done by members of the great family of Nestorian scholars that bore the name of Bukht-Isho'. It produced no less than seven generations of distinguished physicians and translators. The last lived at Bagdad into the second half of the eleventh century but their most active period was 750–850. The skill of the physicians of the Bukht-Isho' family and their school induced the Caliphs to encourage the spread of Greek medical knowledge in Arabic dress.

The translators had command of Greek, Syriac and Arabic and often also of Persian. Most wrote for preference in Syriac but very many works in Arabic were produced by the venerable Yuhannā ibn Māsawaih,* the John Mesue of the Latins (d. 857). He was a pupil of Gabriel b. Buhkt–Isho', the medical adviser to the Commander of the Faithful of romance, Hārūn ar-Rashīd, fifth Abbasid Caliph (reigned 786–809).

John Mesue had many successors, and in time Arabic thus came to replace Syriac for scientific and medical works. Just as 750–850 was the century of translation from Greek into Syriac at Gondisapur, so 850 to 950 was the century of translation into Arabic at Bagdad.

The seventh Abbasid, Caliph Al-Ma'mūn (reigned 813–33), instituted at Bagdad a regular school for translation equipped with a library. Hunain ibn Ishāq (809–77), a philosophical and erudite Nestorian, was the dominating figure. He passed his life in Bagdad, serving nine caliphs, exhibiting phenomenal activity, and translating into Arabic almost the whole immense corpus of Galenic writings. His predilection for the scholastic turn in Galen's theories did much to give that writer his supreme position in the Middle Ages both in the East and in the West.

For anatomical terms for blood vessels and bones, these early translators could have drawn on a singularly copious nomenclature in Arabic derived not only from human beings but also from camels and horses. Yet they went to this source but little, nor is it difficult to see why. The Arabic vocabulary had not yet been marshalled by philologists and these translators were ignorant of many of the existing words known only to Bedouins. Doubtless, too, they found some of the popular terms insufficiently precise. Arabic native dictionaries, even when they became available later, often gave several mutually exclusive meanings to anatomical terms. The material of such dictionaries was, in fact, collected by men learned in literature but ignorant of anatomy. The terms were often culled from poetry which conveyed the sense but vaguely, and represented no real anatomical tradition. Thus the translators into Arabic often neglected native words for others which they themselves formed on Greek models. Not seldom they would transliterate a Greek term with Arabic characters, mutilating the form in the process.

Matters would be comparatively easy if we could trace all these foreign anatomical Arabic terms to their equivalents in Greek medical literature. Unfortunately by no means all the Greek words found in Arabic are known to us from extant Greek sources. (See, for example, note 42 on cephalic and basilic veins.) In some cases it seems tolerably clear that the translators had drawn on the current Byzantine medical terminology, of which our knowledge is but rudimentary.

* Properly *Māsōye*, a Persian diminutive of '*Abd-al-Masīh* ', ' servant of the Messiah '.

The first to compose an original work on medicine in Arabic was probably ‘Alī ibn Rabbān al-Ṭabarī about 850. He was a convert to Islam from Christianity, doubtless of the Nestorian form. It may safely be presumed that he knew Syriac. He was followed by a long succession of writers. Outstanding among them were Muhammad ibn Zakariyyā al-Rāzī of Persia and Bagdad (*Rhazes*, 865–925), ‘Alī ibn ‘Abbās al-Majūsī of Bagdad (*Haly Abbas*, died 994) and Husain ibn ‘Abdallāh Ibn Sīnā of Bukhara and Isfahan (*Avicenna*, 980–1037), all three of Persian origin. In anatomical terminology these added little to the work of the translators into Arabic.

3. *Passage of Arabic Anatomical Nomenclature to Latin West.*

The Western world became first acquainted with Arabic medicine through the Latin paraphrases of Constantine the African (d. 1087). He was born a Moslem or possibly a Jew and probably at or near Kairouan in Tunisia. From there he went to Sicily and thence to Salerno, where he was converted to Christianity. The last ten years of his life were spent at the monastery of Monte Cassino translating Arabic medical works into Latin. His writings were widely studied in the twelfth and thirteenth centuries and continued to be read until the time of Vesalius.

The contact between Orient and Occident produced by the Crusades bore very meagre fruit in the form of learning. Nevertheless it yielded one work important for our purpose. Early in the twelfth century the Pisans established a trading colony at Antioch. There in 1127 one Stephen made a translation into Latin of a work of al-Majūsī (*Haly Abbas*), his *KITĀB AL-MALAKĪ*, the *Liber regius* of the Latins. This was more exact than the version of the same work under another name produced by Constantine. Stephen the Pisan—sometimes called Stephen of Antioch—like Constantine the African, adopted the method of substituting established or newly formed Latin anatomical terms for those encountered in the Arabic text.

A quite different technique was adopted by the greatest of all the Latin translators into Latin from Arabic, Gerard of Cremona (1114–87). He worked in Spain at Toledo, and had there the constant collaboration of Arabic-speaking native Spanish Christians and Jews. They formed a whole school of assistants. Among the eighty-seven works thus rendered by Gerard (or his school) many were on medical topics. Several attained a wide popularity, notably the *KITĀB AL-MANSŪRĪ* of Rhazes, known as the *Almansor* to the Latins, and, above all, the great *Canon* (*AL-QĀNŪN FĪ 'L-ṬIBB*) of Avicenna. The influence of the last work on western medicine of the middle ages and renaissance can hardly be over-estimated.

The literary milieu of Gerard at Arabic-speaking Toledo contrasts with that of Constantine in the Latin monastery of Monte Cassino and with that of Stephen at fortress-like Antioch. In Spain there was an ancient and well-established tradition of learning in Arabic medicine. The recipients of this tradition were accustomed to employ Arabic terms colloquially in a technical manner, much as Latin words are used in present-day medical language. There are many of these transliterated Arabic words in Gerard's translations into Latin. (See for example note 40 on *guidez*.) From translations into Latin these words passed to the current medical jargon of Europe, and some of them entered the vernaculars. Many became corrupted in the passage and many were again modified after they had so passed.

We can often follow the first stage in these transformations from the classical Arabic to the Latin form selected by Gerard and other translators. For this process of detection there are several fortunate adventitious aids. Thus certain Arabic sounds are unknown in Latin. Again, Spanish-Arabic pronunciation had its own peculiarities. Further, many terms passed through the Hebrew of Jewish writers or interpreters, whereby the vowels of some words were recast according to Hebrew patterns. The strange-sounding words were, however, naturally further corrupted in manuscript tradition and have sometimes reached the *Tabulae* of Vesalius in extremely curious transformations. (See, for example, note 46 on *median* and note 54 on *salvatella*.)

We may say therefore that the Arabic element in medieval Latin anatomical terminology is two-fold. Firstly there are Latin loan-translations of Arabic terms, which may or may not ultimately go back to Greek originals. Secondly, and more numerous, are transliterated Arabic words more or less modified, often by Hebrew transmitters. The *Tabulae* provides numerous examples of both classes.*

§ 4. *The Hebrew Element in the Renaissance Anatomical Vocabulary.*

There were medical writings in the Hebrew language during the Dark Ages. Specimens survive in the works of Asaph (8th cent.) and Donnolo (10th cent.). This literature had no influence on the material that we are treating. We can disregard these early works and turn to consider a very definite group of later medieval works in Hebrew translated from Arabic. They were prepared during the same period as that in which the Latin translations were made from the Arabic (11th to 13th centuries), mostly in Italy and to a much less extent in Spain and Provence.

This order of the places of origin of Hebrew translations from Arabic—placing Italy first, Provence next, and Spain last—may cause surprise, since for Latin translations from Arabic Spain easily takes first place. But it must be remembered that most Spanish Jews at this period could speak and read Arabic and that Provencal Jews could often do so. Such persons had, therefore, no need for translations of Arabic works into Hebrew. The reputation of Jewish doctors of the time was based on their knowledge of Arabic medicine. But Italian Jews could seldom read Arabic† and few had facility in Latin. Translations into Hebrew were therefore of great importance to Italian Jewish doctors. These versions provided the portal into the Arabic medical system which, despite its faults, was

---

* For any study of Arabic medicine the work of F. Wüstenfeld, *Geschichte der arabischen Aerzte und Naturforscher*, Göttingen, 1840, is still unreplaced. For the passage of Medicine from Greek into Arabic the most important works are M. Steinschneider, *Die griechischen Aerzte in arabischen Uebersetzungen* in the 'Archiv für pathologische Anatomie etc.', Bd. XXIV, pp. 115, 268, 455, Berlin, 1891; M. Meyerhof, *New Light on Hunain Ibn Ishaq and his Period*, 'Isis', VIII (1926), pp. 685–724; and M. Meyerhof, *Von Alexandrien nach Baghdad* in the 'Sitzungsberichte der Preussischen Akademie der Wissenschaften', Berlin, 1930.

† Unless they came from Sicily or were immigrants from Spain.

far superior to the debased classical Graeco-Latin tradition which it was soon to displace among the Latins.

The Italian Jewish physicians, who used these Hebrew texts, enjoyed great popularity for about three centuries. Their decline, however, had set in some time before the arrival of Vesalius. It was due partly to legal restrictions on their practice, partly to their exclusion from the universities where the newer scientific methods were just beginning, partly to the revival of Greek learning which was displacing ' Arabian ' science. The great centuries of Jewish medicine were from the eleventh to the fourteenth. By the fifteenth century it was in decay. By the sixteenth century medical books in Hebrew had lost their importance. They were neglected and forgotten.

Thus it is not remarkable that so few medical manuscripts in Hebrew of the fifteenth century or earlier have survived, despite the great number that once existed. The only Hebrew medical text of this important literature that was printed was the *Canon* of Avicenna. It was produced in 1491 at Naples by Azriel ben Joseph, who was a native of Italy but of Franconian origin.

This now rare class of Hebrew medical texts which we have been discussing must be distinguished from the very numerous class of medical MSS in the Arabic language written in Hebrew characters. Moreover, it has to be distinguished from another class of Hebrew medical literature. As the centuries wore on, and as Jewish literature passed to its nadir in the sixteenth and later centuries, Hebrew versions of the great Arabic medical works were replaced by popular medical compendia in Hebrew. But we have here to consider neither Arabic medical literature in Hebrew script nor the lower grade of popular Hebrew medicine. We are concerned only with medieval Hebrew translations of classical Arabic medical works, exemplified primarily by Avicenna's *Canon*.

The Hebrew text of Avicenna printed at Naples by Azriel ben Joseph is based to a great extent on the text of three separate Hebrew translations of the work by three Italian translators. These men are Nathan ha-Meathi and Zerahiah Graciano, both of whom lived in the 13th century, and Joshua Lorci of the 14th. Graciano and Lorci worked on only the first two of the five books of the *Canon* but all three translations include the anatomical section with which we are concerned. In this section the printed Hebrew text follows mainly the terminology of Graciano, but it also contains many Hebrew words not in the MSS of Graciano, Meathi or Lorci, who often use transliterated Arabic words instead of Hebrew words. These additional anatomical terms were , introduced by the fifteenth-century editors. It is improbable that Azriel ben Joseph, the fifteenth-century printer of the volume, could draw on Arabic. The colophon mentions only two proof-readers, and they, being from the Rhineland, could not be expected to know Arabic. These new Hebrew terms represent merely purist improvements, not a new effort of translation from the original.

The printed Hebrew Avicenna was used by Lazarus de Frigeis, the Jewish friend of Vesalius, as a source of the Hebrew names of the bones as given in the *Fabrica*. The printed Hebrew Avicenna is not, however, a source for the Semitic terms in the *Tabulae*. For the Hebrew and Arabic terminology of the *Tabulae*, Vesalius depended on some Jewish adviser (or advisers) other than Lazarus. The sources of this man (or men) do not correspond to any Hebrew medical works known to us. His vocabulary is very different from that of Lazarus and of the *Fabrica* and contains several non-technical Arabic words. He exhibits many erroneous substitutions and mispronunciations. His usage of Arabic was corrupt, colloquial and highly specialized. It was probably the current medical jargon of the Italian Jewish doctors of the time. They were a depressed and decaying class. Their hold on the Arabic anatomical tradition was at best residual and precarious, and might perhaps be compared to the last stages of legal French in England.

The *Tabulae*, then (unlike the *Fabrica*), is not influenced by the printed Hebrew Avicenna. It is quite clear that in 1538, when Vesalius published the *Tabulae*, he and his adviser were unaware of the existence of that work. We may thus take it that Lazarus de Frigeis had no part in the composition of the *Tabulae*. Presumably, therefore, Vesalius had not met him in 1538.* In any event, the Semitic terms in the *Tabulae* of 1538 suggest a very different scientific atmosphere from that of the Semitic terms in the *Fabrica* in 1543.

The Latin transcriptions of Hebrew and Arabic words in the *Tabulae* contain some errors. These have a special interest when they show how these Hebrew and hebraized Arabic words were pronounced and used in Italy. They certainly indicate how little the original Arabic forms were remembered by the Italian Jewish doctors of the time. They provide, in fact, a sample of what was something approaching a medical language or jargon now lost beyond recall.

Though some Arabic words in the *Tabulae* (and elsewhere in medical literature) betray by their hebraized form that they have passed through Jewish intermediaries before they came into Latin, very few original Hebrew terms reached non-Jewish anatomical usage. One is the very curious term *gafherua* (note 235) for the innominate bone, while another, with a very long history of a different order, is *canna*, the modern ' cane ', for one of the long bones of forearm or leg (256).

To sum up, then, the Semitic words quoted by Vesalius in the *Tabulae* are representative of contemporary oral usage of obscure sources, Arabic, Hebrew and Romance. They give a glimpse of the final stages of a separate medical tradition expressed in what was almost a separate language.†

§ 5. *Did Vesalius know Arabic and Hebrew ?*

The knowledge of Hebrew possessed by Vesalius did not go beyond the alphabet. Certain errors in the printing of the Hebrew script in both works suggest that he had hardly even this acquaintance with the language. Furthermore, Vesalius seems scarcely to have realized that Hebrew and Arabic are completely separate languages, about as far from each other as English is, for example, from German. Such ignorance was, however, much more excusable then than now. In the sixteenth century the Arabic language was usually written in Hebrew script in

* A hypothesis which seems to us less tenable is that Lazarus helped Vesalius for both *Tabulae* and *Fabrica*, but that in 1537, when the *Tabulae* was going through the press, he knew nothing of the Hebrew Avicenna, either printed or in MS.

† The bibliography of Hebrew medicine is very comprehensively treated by M. Steinschneider, *Die hebräischen Uebersetzungen des Mittelalters und die Juden als Dolmetscher*, Berlin, 1893. See also F. H. Garrison, *Bibliographie der Arbeiten Moritz Steinschneiders zur Geschichte der Medizin*, in the ' Archiv für Geschichte der Medizin ', Leipzig, 1932. The subject is reviewed by H. Friedenwald, *The Jews and Medicine*, two vols., Baltimore, 1944, pp. 146–84.

Europe, where Jews were almost the sole repositories of Arabic learning. But that Vesalius was completely ignorant of both Arabic and Hebrew is a conclusion that seems inevitable after a study of the treatment of Semitic terms in the *Tabulae*.

It is thus surprising to find that in 1537, before the printing of the *Tabulae*, Vesalius was hinting that he had a working knowledge of Arabic. Had this precious possession really been his, it would have been truly a matter of note. In those days a knowledge of Arabic was an excessively rare accomplishment, even among the great philological students. Yet in the prefatory letter (dated 1 February, 1537) to his *Paraphrasis in nonum librum Rhazae Medici Arabis clarissimi ad Regem Almansorem*, Vesalius writes to Florenas :—

" I have started to revise the translation of Rhazes to enable the author himself to reach men's hands cleansed of all barbarian names of medicaments. I did not always attempt to translate word for word, though perhaps a translator should do so. Rather did I prefer to paraphrase . . . . often explaining what seemed most obscure. I fear therefore my labours will not be safe from the bites of fault-seekers, who do not wish anything cleansed from its barbarity, and do not value anything which they themselves have not learnt. . . . Perchance those haters of the success of others stand in awe of your exceptional knowledge of Greek and Arabic medical literature."

It was part of the boastful manner of writers of the time, and almost one of their conventions, to suggest the possession of a store of esoteric erudition. But in this passage Vesalius passes rather beyond mere convention. The man who intends to deceive in this way is certainly not above some of the less lovable human failings. The most enthusiastic admirer of Vesalius will hardly claim him as a shining example of rigid veracity.

§ 6. *General Character of Medieval Hebrew.*

Hebrew anatomical terms reflect, on a small scale, the peculiar character of medieval Hebrew. It was a purely literary idiom used by people who spoke many different languages, both European and Asiatic, in their daily life. Thus its history bears some analogy to that of medieval Latin, though in many essential respects it is utterly different from it. We may consider some of these differences.

Latin in the middle ages passed through periods when, owing to lack of education, elements of its daughter languages were allowed to invade it. And it passed through other periods when men sought to achieve the style of the great classical authors. Yet there was always only one generally recognized source of the pure language, namely the works of the great writers of classical antiquity. Medieval Hebrew, on the other hand, began with not one but two models. These were 'Biblical Hebrew', as used in writings from the twelfth century B.C. to the second B.C., and 'Mishnaic Hebrew', which was in general use from then until about 300 A.D. The two are almost distinct languages. Moreover, from about 300 to 700 A.D. hardly any Hebrew, either Biblical or Mishnaic, was written. It had been replaced, both in oral and literary use, by various forms of Aramaic, a language sufficiently related to both types of Hebrew for mutual influence to be exerted. From 700 onwards there was a series of movements for the revival of the Hebrew language (in writing only), each affecting only part of the Jewish people and usually only certain genres of literature.

It is the Hebrew scientific literature of the later middle ages in S.W. Europe which chiefly concerns us here. The linguistic basis of this was the confluence in the twelfth century of a predominantly Mishnaic strain from north-west Europe and a predominantly Biblical strain from Spain. In both these types there had been many innovations by that time. The idiom that resulted from the confluence was, in effect, yet another new language which developed its own living tradition. Writers of it seldom referred back to Biblical or Mishnaic literature as standards of accuracy. In addition to all this, the translations from Arabic brought in a large and all-pervading Arabic element. Further, the habits of European languages penetrated even into the syntax. Thus the language developed along its own very peculiar lines.

The relationship of this type of medieval Hebrew to Biblical Hebrew can therefore hardly be compared to that of medieval to classical Latin. It would give a clearer idea of this relationship to say that the difference of medieval Hebrew from the language of the Bible is not unlike that of modern English from Anglo-Saxon. There is this, however, to be remembered, that the Hebrew Bible, in contrast to Anglo-Saxon works, was in the hands of every Jew. In the form of conscious or unconscious quotation Biblical Hebrew thus exercised constant and considerable influence on every form of medieval Hebrew.

§ 7. *Biblical and Mishnaic Terms in the 'Tabulae'.*

With regard to the anatomical terms of everyday life, the difference between Biblical and Mishnaic Hebrew is not very great. Such little anatomy as can be found in Mishnaic literature refers to slaughter of animals and ritual purity. It was only natural that in the discussion of these the Mishna itself used the terminology of the Biblical laws on which it was based, even though different words were current in colloquial Mishnaic. Such Biblical words, used in exactly the same way as they are in the Bible, are *e.g.* SELA' = rib (note 230), SHOQ = lower leg (255), 'AQEBH = heel (271), KATHEPH = shoulder-blade (305), 'ASEH = coccyx (320), and 'ESEM = bone (frequent).

The Biblical words in the *Tabulae* are not always, however, given in a grammatically correct form. Thus SHINNAYIM = teeth (99) is a good Biblical word, but in the Bible it is always feminine. In the derived words following SHINNAYIM, such as MEHATTEKHIM (101), KALBIYYIM (102), and TOHANIM (103), we see that the word is treated as a masculine noun. This was a regular practice in medieval Hebrew, where only words recognizable as such by their external form were treated as feminines. 'AQEBH = heel (271) is transcribed *aekef*, which shows that in medieval usage it had changed its first vowel, probably by contamination with the frequently used 'EQEBH = consequence.

Mishnaic words in the *Tabulae* are : WERIDH = a vein (18), HULYAH = vertebra (313), ARKUBBAH = knee (164), all three probably originally from Aramaic. Pure Hebrew words, which occur in the Mishna, but not in the Bible, are SHEDHRAH = spine (309), QANEH in the sense of forearm (138 : in the Bible it means only reed), MOAH = brain (193 : in the Bible = marrow).

Since the terms found in the Biblical and Mishnaic literatures are not sufficient for scientific anatomy, the Hebrew translators and writers of the middle ages cast around for additional words.   Some could be provided by giving a slightly different use to an old word.   Thus the old translators of Avicenna into Hebrew (p. lxxvi), seeking a word for the canine teeth, hit upon the Biblical *hapax legomenon* MALTĔ- 'OTH KEPHIRIM = teeth of the lions (*Psalm* LVIII. 7).   In employing this word for the canine teeth, they merely followed the Jewish Bible commentators (*cf.* 102).   ḤAZEH = sternum (121) is a term from the biblical law of sacrifice, where it denoted the centre-piece of the animal's breast (in the LXX **stēthynion**), which was the priest's share (*Leviticus* VII. 30 *seq.*).   ŽEROA' = upper arm (131) is in the Bible mostly = forearm.   'OREPH = occiput (210) is in the Bible the nape of the neck, by which one can seize a person (*Genesis* XLIX. 8 ; in the LXX **nōtos** = back).   The sense of 'occiput' may be derived from the fact that it is in *Jeremiah* XVIII. 17 contrasted with the face, and PANAH 'OREPH means to turn one's back.   QANEH = reed is used for tibia and fibula (256, 262), because these are called reeds in Arabic.   The remoter history of this word is of very great interest.   The word ḤEṢ = arrow is used as translation of the Arabic for the sagittal suture (285).   PAḤADH, a word of uncertain meaning, occurs but once in the Bible and is explained by the Jewish commentators as testicle.   In the *Tabulae* (340) it is used for thigh, because of its similarity with Arabic *FAKHIDH*.*   In HA-NABHUBH = *vena cava* (37) a Biblical adjective is given a new meaning by the all-too literal treatment of a Greek term.   In HA-NIRDAMIM = carotid (81) the translation is even clumsier.   The 'sleep-causing' arteries have become the 'sleeping ones'.   The mistake is understandable since the adjective in Arabic does not indicate in what connection the arteries stand to slumber, but the term was coined unintelligently.

Mishnaic terms, too, were given some novel uses.   Thus MASREQ = comb (156) came to be used for the metacarpus in imitation of Greek-Arabic usage. QĔDHERAH = bowl, entered into the term QEDHERATH HA-MOAH = bowl of the brain = skull, an image of Romance origin.

§ 8. *Medieval Hebrew Terms in the 'Tabulae'.*

New derivations from Hebrew material are very rare in the *Tabulae*.   They are restricted to a few adjectives formed from nouns by the 'nisbe' form in which -*ī* is attached to the stem.   Thus we get KALBI = *caninus* (102) from KELEBH = dog, and KITHRI = *coronalis* (279) from KETHER = crown.   The feeling for the correct grammatical handling of this formation seems to have been lost, for KITHRI is transcribed *cethari*, a grammatical impossibility.

We have seen that some new terms were imitations of Arabic words, of the type called by the linguists loan-translation or *calque*.   But further Arabic words were introduced merely transliterated into Hebrew characters.   It is difficult to understand why this should be done in cases where good Hebrew terms were available or where the Arabic could have been imitated.   Possibly Jewish physicians liked to grace their books with mysterious Arabic words, as did their Christian contemporaries (see p. lxxv f.).   But there was some antagonism to this tendency, for in the printed Hebrew Avicenna a number of these Arabic terms are replaced by Hebrew (p. lxxvi), *e.g.* TARQUWA = clavicle by SHEKHEM (*cf.* 107) though the latter word is less exact.

---

* The Arabic word, hebraized according to the system set out in the next section, would have the same spelling as the Hebrew.

In transliterating Arabic words into Hebrew a difficulty arose in that Arabic has more sounds than Hebrew. It may be useful to set down here the method by which such sounds were usually rendered in the Middle Ages.

Arabic:	TH	J	KH	DH	D*	Z	GH	ث	ج خ ز	ض ظ	غ
Hebrew:	T	G	Ḥ	D	Ṣ	Ṣ	G	ת	נ ח ד	צ צ	ג

All other consonants were rendered in Hebrew by letters which we have transcribed here by the same signs in the two languages.

The correspondences given above are only those found in loan-words. They are neither identical with etymological equations nor with the spelling by which Arabic was normally written in Hebrew characters.† In oral transmission some further changes were made, as when ZORQI = $ZAURAQ\bar{I}$ = *naviculare* (176) becomes in our text ṢORQI. The Arabic definite article AL- was sometimes incorporated into the Hebrew word, as in AL'AGHAZ = AL-'$AJUZ$ = buttocks (326), in ALZARGĒPHĀ = AL-$HARQAFA$ = ilium (238), and in ALṬA'IGHĀ = *theca* = acetabulum (244). This incorporation of AL- is frequent in those Arabic words which passed into Latin, and is well known in the common nouns alcohol, alcove, algebra, alembic etc.

Arabic words were not always taken over in the singular form. Thus we have in our text ZINĀD = bone of the forearm (143), which is a plural of $ZAND$. Among words that passed into the usage of the Latins we find *aresfatu* = AL-$RADAF\bar{A}T$ = patella (165) and *althavorat*, probably AL-$FAWW\bar{A}R\bar{A}T$ = groove of the groin (246), both plurals. In none of these cases can any reason be discerned for choosing the plural.

Arabic words having no obvious etymological background, were naturally exposed to greater hazards than Hebrew words in the process of transmission. They suffered from a tendency to reduction to common Hebrew vowel-patterns, as in the case of $RUSGH$ = metacarpus, which became RESEGH (152) = medieval Latin *rasceta* and '$AJUZ$ = sacrum, which became AL'AGHAZ (326). Both of these are reduced to the common Hebrew ' segolate ' pattern with two E vowels or, in contact with guttural consonants,‡ with two A vowels. Others were changed

---

* The reader may be surprised to find Arabic D rendered by Hebrew Ṣ. In fact, the Arabic sound conventionally so transliterated was originally a sibilant, and in the Middle Ages probably pronounced as emphatic DH. In Persian and Turkish loan-words from Arabic it is pronounced Z. The current transliteration as D is based on the form the sound has taken in most types of modern colloquial Arabic.

†

Arabic:	TH	J	KH	DH	D	Z	GH
Etymological correspondent in Hebrew:	SH	G	Ḥ	Z	Ṣ	Ṣ	y
Usual way of spelling in Judaeo-Arabic MSS.:	T with dot	G with dot	K with dot	D with dot	Ṣ with dot	Ṭ with dot	G

Sometimes the spelling of Arabic J and GH is reversed. The dots are often carelessly omitted in MSS.

‡ i.e. 'Aleph, 'ayin, Ḥ, H.

according to some popular etymology, as *QAṬN* = loins, which became QAṬON (332), as if it was the Hebrew word for ' small '. Forms that were as alien to Hebrew speech-habits as *TARQUWA* = clavicle (107) were made at least pronounceable by recasting, in this case, into TARQUHA. This is an impossible form for a noun in Hebrew, but would be possible as a third person plural of a (non-existent) verb in the perfect tense with the object-suffix of the third sing. fem.

§ 9. *Hebraic Corruptions in the ' Tabulae '.*

Beyond such hardly avoidable changes, Arabic words suffered by sheer corruption. It is interesting that the cases of profound corruption all go back to misreadings, not to the vicissitudes of oral tradition. TARDI = tarsus (180) might become in oral tradition TARDI, as the two kinds of T were not distinguished in the Hebrew pronunciation of Italian Jews, but the substitution of TARDI for NARDI = backgammon stone, and the other corrupt form, BARDI, found in the printed Hebrew Avicenna, can be explained only from the stumbling blocks of Arabic writing, where *B, T, TH, N* and *Y* all look the same if the scribe be not meticulous in dotting his letters. Again, the form KARDI, which appears in the *Fabrica*, is due to the tiresome similarity of B and K in Hebrew print. Yet such corruptions had a lasting influence even on Latin terminology. Thus BARDI gave rise to a Latin *grandinosum* (182) because BARADH = hail = *grando*. Men who made mistakes of this kind were surely working with their eyes and not their ears.

The inability of Hebrew script to distinguish between Arabic *Ḥ* and *KH* led people to understand *AL-KHANJARĪ* = dagger bone = xiphoid cartilage as if it were *AL-ḤANJARĪ* = throat bone. Graciano promptly rendered this spurious term into Hebrew as 'EṢEM GĔRONI, and the Latin development of this appears in our text as *epiglottalis* (129). This is, in fact, the common medieval term for the xiphoid and has puzzled many modern historians. Some medieval, and even some modern, writers have laboured faithfully to discover the similarities between the xiphoid cartilage of the sternum and the Adam's apple. A whole train of theory was thus set in movement by a single missing dot !

Similarly the lack of signs for vowel length in Hebrew allowed Graciano to mistake *AL-'AJUZ* = *os sacrum* for *AL-'AJŪZ* = the old woman. This was hebraized as HA-ZAQEN = the old man (326). The similarity of W and Y in Hebrew writing similarly caused *AL-'AZM AL-MUQAWWAS* = the curved bone = *naviculare* to be misread as *AL-'AZM AL-MUQAYYAS* = the equal bone, = 'EṢEM HA-SHAWEH (176), a senseless name given to the bone in the printed Hebrew Avicenna.

Corruptions did not have to make sense in order to find a place in the tradition which reached Vesalius. Thus the plain Hebrew HA-GARON = the throat (315) was corrupted into MEGARON (H and M being vaguely similar.) This form was so transliterated into Latin characters by Vesalius' assistant. Whole chains of misreadings, mistranslations and misdivisions are responsible for the creations ALZEGEM ḤARṬUM = acromion (113) and SHĔḤUSI = temporal suture (294).

Such changes and corruptions as we have discussed sometimes make it possible to prove that a latinized Arabic term must have passed through Hebrew. In some cases this may be done with assurance even though the Hebrew link be missing.

Thus *aseth* = brachium (133) from *'ADUD* has changed its vowels to conform to the segolate pattern (see p. lxxxi). True to this the word appears in the *Fabrica* transliterated into Hebrew characters and with *hasad* as the pronunciation.

§ 10. *General Character of the Arabic Vocabulary in the 'Tabulae'.*

The Arabic terms taken over and imitated in Hebrew and Latin had been themselves the product of a long and involved development. This process has not yet been traced in detail by philologists. The history of the Arabic language is still insufficiently known. Thus many Arabic anatomical terms have no known etymology. We are ignorant of the original meaning of words like *USAILIM* = salvatella (54), *SĀFIN* = saphenous, *FAWWĀRA* = groove of the groin (246).* We suspect that they were once appellatives derived from some general root and denoting some special quality of the part, but as with so many Arabic words, they are remnants of extinct roots of unknown significance.

Other Arabic terms are easily analysed: *MĀDHIYĀNI* = *mediana* (49) means the two freely flowing ones, *AKHAL* = *nigra* for the same vein (50) means the blue-black one, *ZAND* = *focile* (143), means fire-stick, etc. Some of these terms betray themselves as butcher's words. Such may be *HABL AL-DHIRĀ'* = rope of the forearm = *funis brachii* (51, *cf.* the note on the origin of Greek **aortē,** 71). From the butcher's trade may derive also such vague terms as *QABĪLA* = any cranial bone (220), *ANQĀ'* = any large bone with marrow in it (239). Among the Latins the meaning of these words obtained precision. Thus *cavilla* became to mean sphenoid bone (220), and *anchae* came to mean haunch-bone (239). Now and again we can discover in some name the echo of ancient medical lore. Thus *SHA'N*, suture (195), latinized as *soonia*, comes from a root meaning to run, in the sense of drip. The noun meant originally tear-duct. It reflects some belief that humours penetrate from the inside of the head through the sutures.†

As Jews of southern Europe turned naturally to Arabic to supply the deficiencies of Hebrew anatomical terminology, so Arabs turned naturally to Greek, which was the ultimate source of their medical tradition. Sometimes they simply transliterated Greek words. These later passed, again in transliteration, into Hebrew. There these words often became even more modified than the native Arabic terms. An interesting example is **aorte,** which became *AWURTĪ*. In Jewish tradition it acquired somehow a medial *i*, and occurs in two versions. One is AWRITAY, found in the Oxford MS of Graciano's Hebrew translation of Avicenna's *Canon*. This was adopted by Stephen of Pisa as *aurithia*, thus suggesting that even he, who worked in Antioch, invoked the help of Jews for interpreting Arabic texts. The second form appears in Gerard of Cremona as *orithi*. Now the Italian Jewish doctors appear to have noticed that this was nothing but the well-known *aorta*. They therefore dropped the *i* again, and made the word more similar to *aorta* by

---

* It is, however, possible that *althavorat* is not connected with *FAWWARA*. In a skeleton drawing of 1497 the *os pectinis* is said to be called by the Arabs *halhatafar* (*cf.* H. E. Sigerist, *The Book of Cirurgia, by Hieronymus Brunschwig*, p. 261). It may be that both 'Latin' words are corruptions of a yet unidentified Arabic form.

† *Cf.* J. M. Peñuela, '*Die Goldene*' *des Ibn al-Munāṣif*, Rome, 1941, pp. 40 (line 8) and 77, and Lane's *Arabic Lexicon*, p. 1491.

prefixing to it the definite article. So we get HA'ORTI (73). By imitation of this form the definite article was attached to the name of the associated vena cava, HA-NABHUBH (37), though the Hebrew names usually appear in the *Tabulae* without the article.

Akrōmion was transcribed as *AKRAM*. But this word was by popular etymology made into *AL-AKHRAM* = the one with perforated nose.* This was in Hebrew script further transformed into ALZEGEM (113) and into other even more corrupt forms.

Zygōma was identified with the old Greek loan-word *ZAUJ* from **zygon**=pair, which also existed in Hebrew as ZUGH. In Hebrew oral usage a distinction seems to have been maintained between *zog* (213) = zygomatic arch = *ZAUJ* and the common word ZUGH = pair.

Besides these transliterations, the Arabs took over a larger number of Greek terms by the process of loan-translation. Among these we may mention *AL-AJWAF* = cava (37), *SUBĀTIYYĀNI* (*subeticae*) = **karotides** (80), a not very happy translation, *MISHT* = **ktenion** = metacarpus (155), *SIMSIMĀNIYYA* (*simenia, sesamina*) = **sesamoeidē** (188).

Occasionally the Arabs substituted a familiar idea for something specifically Greek, as when they called the ' sigma of the shoulder-blade ' the ' Ayin of the shoulder-blade ' (118). This was a rather doubtful improvement, for while the Greek term was a comparison to the old sigma = C, the Arabs were thinking of the later form of the letter = Σ, which is very similar to Arabic 'Ayn. But '*AYN* = eye, and this coincidence caused a series of efforts to explain how the coracoid (*oculus scapulae*) can resemble an eye.

In other cases the Arabic transcription survived in its original form. The lambdoid suture remained *AL-LAMĪ*, and necessitated a note in every Arabic medical book that it resembled the Greek **lambda**, not the Arabic letter *LĀM*, to which it bears no similarity. In the tradition from which Vesalius drew, a happy improvement had been made by substituting the Hebrew letter TETH, pointed at the bottom, for the unknown Greek sign (282).

One interesting feature about Arabic tradition is that it has terms that are Greek in appearance but have not survived in any Greek work. Thus we have *QĪFĀL* = cephalica and *BĀSĪLIQ* = basilica (42), which can hardly be explained save by reference to Greek **kephalos** = head, and **basileus** = king. Yet these terms are unknown in Greek as applied to veins. Perhaps there was some local Asiatic-Greek tradition now lost which produced this association. The two words appear first in a translation into Arabic of the early 10th century, which came from the school of Ḥunain b. Isḥāq. That school embodied both the literary learning based on the classical Greek writers and the living oral tradition of generations of Nestorian physicians. But the story behind these words is lost.†

In a few cases the Arabic writers enriched the medical vocabulary by naming parts that had no names in the preceding Greek or Syriac. These innovations

* But *cf.* also the term *AL-AKHRAMĀNI*, denoting ' two cleft bones at the extremity of the interior of the mouth ' (='premaxillæ'), to which this may have been assimilated.

† A. Macalister has a valuable note on the names of the veins of the arm in the *Journal of Anatomy and Physiology*, Vol. XXXIII, New Series, XIII, London, 1899, p. 343.

were mainly of little significance, often merely completing what the Greeks had begun. Thus Avicenna gave currency to the idea of ' true ribs ' as counterpart to Galen's ' false ribs ' (232), no very startling invention. The sagittal suture had no specially significant name in Galen, being called merely the ' straight middle one ' (284). The Arabic writers, seeing the scheme of the sutures as a bow, with the sagittal suture as the arrow across it, called it $AL\text{-}SAHM\bar{I}$ (285) = the arrow-like, of which *sagittalis* is a translation. *Vena porta* again is ultimately due to the Arabic writers (18). Finding difficulty in translating Galen's **stelechiaia** = trunk-vein (17), they sought a term which would indicate more directly the place and nature of the vessel. The Arabic writers also named the innominate bones, though rather poorly, by giving a wider meaning to $AL\text{-}'\bar{A}NA$ = pubes. In this they were imitated by the Hebrews with the clumsy GABH HA-'ERWAH = back of the genitals (235). This, in the form *gafherua*, became one of the few originally Hebrew words to pass into Latin usage.

§ 11. *Hebraized Romance Terms in the ' Tabulae '.*

The most recent stratum of the Semitic vocabulary of the *Tabulae* is formed by those Hebrew terms that are adaptations of Romance words. Thus KAL-BIYYIM = canine teeth (102) is obviously a loan-translation of the Italian or Latin term, being derived from KELEBH = dog, as *caninus* is from *canis*. It is used as a familiar term by Graciano at the end of the thirteenth century. Indeed, it was so current that its origin seems to have been forgotten, and it was changed to QALBIYYOTH, perhaps intended to mean ' heart-teeth ' (Arabic $QALB$ = heart). It may be surmised that such a word was not invented by medical men translating from the Arabic, but was one of the more or less jocular hebraizations used in the speech of Italian Jews.

An example of the jocular element is encountered in the *Tabulae*. For the normal GULGOLETH = skull (cf. *Golgotha*) of all Avicenna translators, we have QĔDHERATH HA-MOAH = calvarium (193). This phrase means ' the bowl of the brain ' = *olla capitis*, and renders the Italian *testa* = skull but originally = earthenware bowl. ' Testa ' is a product of the rude sense of humour of Roman soldiers. This rather poor joke so appealed to Italian Jews that they imitated it in Hebrew. The phrase could not have been invented by one who had the Bible in mind, for GULLAH = bowl is actually used metaphorically for the skull (*Ecclesiastes* XII. 5). Thus any learned origin of the term may be excluded. But as with QALBIYYOTH, the origin of the word was (perhaps intentionally) obscured by a slight change. In the *Tabulae* it appears as QADHRUTH HA-MOAH. Literally this means ' blackness of the brain ' (cf. *Isaiah* L. 3). The phrase can perhaps be justified as = ' cover of the brain ' by taking a clue from *Ezekiel* XXXII. 7, where HIQDARTI = I darkened, is in parallelism with KISSETHI = I covered.

MAGHEN HA-ARKUBBAH = patella, literally ' shield of the knee ' (164), is perhaps also formed after some Romance pattern, though we have not discovered the model. In Arabic it is called ' eye of the knee ', in Italian ' little wheel (or millstone) of the knee ', in German ' disk of the knee ', etc. In Hebrew a suitable term was not available.

Romance influence seems also to have been at work in substituting KITHRĬ = coronalis (279) for the KĔLILI = garland-suture = **stephaniaia** of the Hebrew Avicenna translations. *Corona*, of course, means both garland and crown, but it was more familiar to Italian Jews in the meaning of crown = KETHER (see also p. lxxx).

The 13th century Hebrew translation of Avicenna by Graciano seems to represent most fully the real usage of contemporary Jewish doctors. In this version many Italian words appear in Hebrew transcription, as MUSHKULO = muscle, etc. It often gives the Italian term transcribed in Hebrew letters side by side with the Hebrew. No such words are found in the *Tabulae*. This may indicate that the tendency of the oral medical tradition was to become more exclusively Semitic, excluding Italian terms so as to be less intelligible to the layman. But the tendency was in keeping with the classicist movement in the Hebrew literature of Italian Jews. Thus we find that Lorci in the 15th century does not follow the Romance tendency of Graciano. The printed Hebrew Avicenna has gone further in its classicism. Not only does it not admit any Romance words, but even replaces Arabic transliterated terms by Hebrew equivalents.

The only Romance word in the Hebrew of the *Tabulae* is not borrowed from Italian, but obtained entrance disguised as Arabic. This is AL-ṬA'IGHĀ = *pyxis coxae* (244). The word has no Arabic etymology. It is not found in any Arabic or Hebrew writings that we have consulted. We believe it to be nothing but a Romance form of *theca* with the Arabic article added. The form *taiga* is nearest to Portuguese. The word may have come to Italy as part of the oral tradition of Iberian Jewish refugees.

VIII. Translation and Commentary.

TABULA I.

To his Master and Patron the Most Eminent and Highly Distinguished Doctor Narcissus Parthenopeus [1], First Physician to His Imperial Majesty [2], Andreas Wesalius of Brussels wisheth health.

(1) Verdun, the family name of Narcissus Vertunus Parthenopeus (1941–155?), is well known in Aragon, whence his grandfather had migrated to Naples. Parthenopeus means Neapolitan. Parthenopea was an ancient classical term for Naples, supposedly linked with the syren Parthenope. Narcissus Verdun early rendered service to the court of Spain, of which the kingdom of Naples had been a dependency since 1501. This was recognised by a substantial reward in 1520. In 1524 the Emperor Charles V named him his *conciliarius, protomedicus* and *protochirurgus*. These were doubtless more titles than offices as are ' Privy Councellor ' or ' King's Physician ' nowadays. In 1525 Narcissus was in Spain where he attended Francis I, then a prisoner. In 1530 he was with the court of the Emperor at Bologna and treated him for quinsy. In 1532 he was in Brussels attending a Neapolitan envoy at the Emperor's court. Vesalius, soon to leave for Paris, was then 18 and may have met him through his father who was the Emperor's apothecary (see p. xiv). Narcissus returned in 1534 to Naples, where he remained till his death. Vesalius mentions him again in 1539 in his Venesection Epistle. The ascription by Vesalius to Narcissus of great and various intellectual powers (see page 3) is part of the convention of dedication. There is some reason to regard it as quite unrelated to fact.

(2) Charles was born in Ghent in 1500. His father was Philip of Burgundy. On Philip's death in 1506 he succeeded to the Netherlands and the County of Burgundy. He was declared of age in 1515, the year after Vesalius was born. He had lived till then in Flanders and may have come in contact with the father of Vesalius. Charles was crowned Roman Emperor at Aix in 1530. His estates included the kingdom of Sicily, Sardinia and Naples, hence the connexion with Narcissus Parthenopeus.

Not long since, most learned Narcissus, when appointed at Padua for the course of the surgical part of Medicine [3], I was discussing the treatment of Inflammation. I had come to explain the views of the divine Hippocrates and of Galen on Revulsion and Derivation [4] and had made incidentally a drawing of the veins, thus displaying what Hippocrates meant by **kat' ixin** [5]. I showed that this was easy enough. (Well do I know how debate on the vein to be opened has roused dissension and dispute even among the learned, some claiming that Hippocrates implied a unity and direct connection of the vessels [with the part affected], others another view.) [6] And this figure of the veins so pleased the professors and students of medicine that they pressed me for a similar delineation of the arteries and nerves.

(3) Vesalius was appointed Professor of Surgery at Padua on 6 December 1537, seventeen months before the issue of the *Tabulae Sex*.

(4) Revulsion and Derivation, two familiar terms of medieval medicine, are based on ancient humoral theory. Revulsion is diminution of a diseased state by removing fluid from another part, so that the diseased humour flows away from the affected part to that from which the fluid is removed. Derivation is diminution of a diseased state by removing fluid direct from the affected part.

(5) **Kat' ixin.** The noun **ixis** implies direction. In medical idiom **kat' ixin** = on the corresponding side, on the same side. It is used with this significance in several earlier works of the Hippocratic Collection as well as in Galenic writings. It is applied by Vesalius to the opening of a vein on the same side as the lesion.

| *Latin* | **Greek** | HEBREW | *ARABIC* |

(6) This passage, being of the nature of digression, we have placed between parentheses.   It reflects
      a fierce and futile controversy of the time that gave rise to a very extensive literature.
      The subject is treated by Vesalius, in the following year, in his work of which the full title is
      Epistola docens venam axillarem dextri cubiti in dolore laterali secandam : et melancholicum
      succum ex venae portae ramis ad sedem pertinentibus purgari :   ' Letter inculcating that
      [a branch of] the axillary vein of the right arm is that which should be opened for pain in
      the side : and that the melancholic humour is purged from the branches of the *Vena porta*
      pertaining to the rectum '.   See notes 19, 32 and 58.

Since the conduct of dissections was part of my duty, and knowing this
kind of drawing to be very useful to those attending the demonstrations, I had
to accede to this request.   Nevertheless I am convinced that it is very hard—
nay, futile and impossible—to obtain real anatomical or therapeutic knowledge
from mere figures or formulae, though no one will deny them to be capital aids to
memory.

And now, since many have tried vainly to copy what I have done [7], I have
sent these drawings themselves to the press.   To them we have annexed others
comprising three representations of my skeleton, which I had set up to the
gratification of the students, rendered from the three standard aspects by the
distinguished contemporary artist, John Stephen [van Calcar] [8].   They will
satisfy those who hold it fitting, fair, profitable, nay essential, to contemplate
the skill and craftsmanship of the Great Artist Himself and to peer into the
' house of the soul ' as Plato calls it [9].

(7) The *Tabulae Sex* belong to the tradition of ' fugitive anatomical sheets ', of which very few have
      survived of a date earlier than 1538.
(8) Representations of articulated skeletons occupied some place in medical teaching even in the
      Middle Ages.   They appear in several early printed medical works.
            Vesalius put together his first skeleton in 1536 at Louvain.   That to which he is referring
      in the text, however, he put together in January 1537 for a public dissection at Padua.
      It was of a youth of about eighteen.   His age is shown both by the state of the epiphyses
      and sutures and also by the MS notes of a pupil who calls himself Vitus Tritonius Athesinus
      (on whom see M. Roth, *Andreas Vesalius Bruxellensis*, Basel, 1892).
(9) *Phaedo*, 82.   A very typical humanist allusion.

Against the different parts we have put their names (though this could not
be done quite fully), nor have we excluded such barbarous terms as the learned
sometimes employ in their books [10].

(10) The elucidation of these ' barbarous terms ' and the tracing of their origins have been the main
      difficulties of the present study.   Vesalius has made a rather half-hearted attempt to place
      the descriptive and nomenclatory notes opposite the parts involved.   Much of the material
      in his notes is, however, rather for the display of learning than for elucidation.   In this
      he was merely following the bad fashion of his time.

As for the accuracy of our work, believe me that no item is here which the
Paduan students have not themselves confirmed at my demonstrations of this
year [11].   And concerning my most learned teachers of Paris and doctors of
Louvain, before whom I have often publicly dissected, for the moment I will be
silent [12].

(11) Despite this statement we shall find many points in these figures and in the notes on them which
      Vesalius had certainly not demonstrated on the human body.   Such is, for example, the
      rete mirabile which he portrays.   This exists only in Ungulates, and its appearance in these
      plates is clearly a ' hangover ' from Galenic or medieval anatomy.   See note 78.   Again
      the distribution of the larger vessels is not human, but taken from an ape, probably a Macaca.

See notes 45, 64, 83, 123. The liver bears no resemblance to that of a man or of any other animal. See note 14. Again, an innominate artery giving rise to both subclavians and carotids is found in monkeys and some other animals, but not in man.

(12) The passage indicates that the relations of Vesalius with his colleagues at Paris and Louvain were already strained.

Guenther, with whom he was still on good terms, had left Paris at this date (1538). Sylvius was still there. Who were the other ' teachers of Paris ' to whom he refers ? Did they include Charles Estienne ? It is to be noted that he does not speak of ' doctors of Louvain ' as among his teachers.

That this venture should be weighed by those best fit to judge, it is headed by the name of the most celebrated among them. Thou [Narcissus] hast incomparable facility in tongues and a unique knowledge of anatomy and of all the disciplines of medicine and of philosophy. Throughout all peoples thou art extolled as outstanding, even among distinguished physicians and men of letters. Nay, even the eminent hast thou impressed by the prudence and integrity of thy mind. Thou art admired by all for thy kindness and graciousness. Charles V himself, Ever Victorious Emperor of the Romans, Eternal Augustus, with his keen insight into character, has entrusted thee with the high duty of safeguarding his health. He has placed thee, in the very prime of thy days, as censor of all doctors and drug-shops of Spain and Naples. He has, moreover, marked thee out, even among the eminent, by many honours and gifts.

Accept then, most honoured Sir, the humble gift of these sheets. Accept them with that humanity which thou didst previously extend to me when thy spirit was peculiarly gracious to me. And should I find it acceptable to thee and to the learned, I hope to add something more considerable thereto. Fare thee well. First day of April in the year of Salvation 1538 (13).

(13) The verbosity of this passage and its flattery of Narcissus are in the worst taste and manner of the time ; we have therefore abbreviated the whole paragraph in our translation. Its complete rendering could only increase the distaste which it must arouse.

The Liver (14), workshop of sanguification (15), receives chyle from stomach and intestines through the *vena porta* (16), which is called **stelechiaia** (17) by the Greeks and WĔRIDH HA-SHO'ER *varidhascoer* (18) by the Arabs, and expels the *succus melancholicus* into the spleen (19).

(14) The liver as drawn by Vesalius is not from a human body nor from any animal type. It consists of five equal splayed lobes, and is the conventional representation of the organ familiar to the student of medieval anatomy. It betrays immediately the influence of the medieval tradition on Vesalius.

(15) In the old physiology the liver is the source of the blood which ebbs and flows from it into the veins. This view is firmly based in Galenic theory.

(16) *Porta* = gate is properly the fissure of the liver. The name of the vein should therefore be *Vena portae* = vein of the gate. Vesalius, like many of his contemporaries, calls this vessel *Vena porta* or even simply *Porta*. This usage is not without reason. Gerard translated the Arabic of Avicenna rightly as *Vena quae vocatur porta* or, as we might say, ' porta vein ' = ' gate vein '. The usage gate = *porta* = **pylai** to designate both the fissure of the liver and the great vein associated with it goes right back to Greek sources. See, for example, Galen, *De placitis Hipp. et Platonis*, VI, 5 (Kühn, V, 549) and Rufus, *De appellatione partium*, I, 179.

(17) **stelechiaia**, Galen, *De anat. admin.* VI, 10 (Kühn, II, 575) and *De loc. aff.* VI, 4 (Kühn, IV, 413). These seem the only places where Galen uses the term. **Stelechos** = haft, shaft, trunk, thus **phleps stelechiaia** = trunk-vein.

Latin	**Greek**	HEBREW	*ARABIC*

(18) *Varidhascoer*=WĔRIDH HA-SHO'ER = lit. vein of the gatekeeper. It will be remembered that **pylōros**=gatekeeper. In Arabic, as in Hebrew, the words for *porta*=*BĀB*=SHA'AR and *pylorus*=*BAWWĀB*=SHO'ER are so similar in script as to have caused confusion. That this is so is evidence that the Jewish helper of Vesalius did not consult the printed Hebrew Avicenna, which distinguishes the two correctly.

(19) In medieval physiology each of the four humours has its proper site or organ : Blood in the heart, Phlegm in the brain, Yellow Bile in the liver and especially the gall bladder, and Black Bile or *Succus melancholicus* in the spleen. All these, and especially Black Bile, needed to be purged from time to time. Some medieval anatomical writers held that Black Bile was voided by a duct from the spleen to the stomach. Vesalius had a theory that the process took place from the haemorrhoidal veins. See O and P below, and notes 6 and 32.

A. *Cavum* (20) or *simum* (21) of Liver. ('Transverse fissure'.)

(20) *Cavum*, a hole or cavity, used in the sixteenth century for the hollow or concave side of any organ, as the liver, stomach, etc.

(21) *Simum*, from **simos**, hollow, was a general anatomical term for the cavity in an organ into which another fits. The correct Latin form is *sinum* or *sinus*, but the false form, with an *m* on the stem, was affected by renaissance anatomists.

B. *Vena porta* (22), the *manus* of the liver (23).

(22) On this term see note 16 above.

(23) *Manus* is in general use in sixteenth century anatomy for any structure which terminates by dividing into parts remotely resembling digits.

C. Small branches on the bladder of the yellow bile (24). ('Cystic veins'.)

(24) The distinction of two kinds of bile was fundamental, the gall bladder being reserved for yellow, the spleen for black bile.

D. To *Pancreas* (25) and to *ecphysis* (26) or *intestinum duodenum* (27). ('Pancreatico-duodenal veins'.)

(25) **Pankreas** = all flesh, term used by Galen. Vesalius confuses it with the mass of lymphatic glands at the root of the mesentery (*Fabrica*, 1543, p. 505, line 7 *et passim*). This latter mass = 'pancreas of Aselli', is especially developed in the Carnivora. In the *Fabrica* Vesalius himself notes this for the dog. The confusion of pancreas with the central abdominal lymphatic mass is made by Aristotle, *Hist. an.* (514 a, 10–15), whence Vesalius perhaps took it.

(26) **Ekphysis** = outgrowth. Not a usual or very natural term for the duodenum. Vesalius took it from Galen's *De anat. admin.* Bk. VI, Ch. 12 (Kühn, II, 578), where the duodenum is regarded as a 'growing out' from the stomach. Galen himself derived it from Herophilus. See *De ven. et art. diss.* Ch. I (Kühn, II, 780).

(27) The word *duodenum*, rendering **dōdekadaktylon**, came into Latin with the translation from Arabic of the *Canon* of Avicenna by Gerard of Cremona (*c.* 1170).

E. To the right of the curvature (*gibbus*) (28) of the stomach. ('Coronary veins'.)

(28) *Gibbus* is a favourite term of renaissance anatomists for a broadly curving surface.

F. To the right of the *fundus* of the stomach and to the upper [that is to the superficial] membrane of the *omentum* (29). (Largely 'pyloric veins'.)

(29) *Omentum* is used in its modern sense by both Pliny and Celsus. The ultimate source of the word is obscure. (See Hyrtl, *Onom. anat.* p. 363.) In the *Fabrica*, Lib. V, Ch. 4, Vesalius seems to regard the *omentum* as much thicker than in fact it is.

G. Main bifurcation of the *porta*.

H. Having passed through the lower [that is, deeper] membrane of the *omentum* and through the *pancreas* they are variously distributed. (Mainly 'splenic vein'.)

I. Into the lower [that is deeper] membrane of the *omentum* on the right side. ('Inferior mesenteric' in part.)

K. Through the *cavum* of the stomach, ending by surrounding its *os* ('Cardiac end') with many branches. ('Left gastro-epiploic vein'.)

L. Into the lower [that is deeper] membrane of the *omentum* in its middle part and it branches first into two and then into many minute veins. ('Inferior mesenteric vein' in part.)

M. Divided into many branches which are implanted in a straight line in the *simum* [(29 A)] of the spleen. By this the impure blood [(30)] is conveyed into the spleen. ('Lienal vein'.)

(29 A) On *simum* see note 21.

(30) Impure blood, *sanguis faeculentus*, here means blood charged with melancholic humour. The passage stresses again the doctrine that Black Bile was purged by the spleen. (See notes 19 and 24.) One of the puzzles of the old physiology was to find a passage through which this would be possible. Several solutions were suggested. The answer of Vesalius was through the haemorrhoidal veins and perhaps also through the stomach. (See O and P of this plate.)

N. Each goes to the left curvature (*gibbus*) of the stomach and afterwards somewhat obscurely to the cardiac opening (*os*) of the stomach. ('Vasa brevia'.)

O. To the left of the *fundus* of the stomach and upper [that is, superficial] membrane of the *omentum*. I think that by this vein no small part of the excrement of the spleen is purged into the stomach. ('Inferior mesenteric vein' in part.)

P. It runs out to the intestines in numerous branches between the membranes of the *mesaraeum* [(31)]. ('Superior mesenteric vein'.) I am not sure whether haemorrhoids arise from this or from [branches of] the *vena cava*, for from both these veins [that is, from *vena porta* and *vena cava*] there are branches to that part. The greater are from the [*vena*] *porta*, and it therefore seems reasonable that the melancholic blood should be purged by the [*vena*] *porta* [(32)].

(31) **Mesaraion**, of Galen, *De anat. admin.* Bk. VIII, Ch. 5 (Kühn, II, 561) is anatomically exactly equivalent to **mesenterion** of Aristotle, *Hist. an.* (495 *b*, 32), and remains so throughout the anatomical literature. The element **araion** = slender. The term survived into twentieth century anatomical phraseology as 'mesaraic', but has now passed out of use.

(32) The subject of the purging of black bile through the haemorrhoidal veins is reviewed by Vesalius in his letter on venesection of 1539. (See notes 6 and 19.)

Organs of generation, above of the Man, below of the Woman. The third figure shows the implantation of the *vasa* carrying down the semen [*deferentia*] [(33)].

(33) *Vasa deferentia* is a modern term. It was, so far as can be traced, introduced by Jacopo Berengario da Carpi in his *Commentaria ... super anatomia Mundini*, Bologna, 1521, and, therefore, in the lifetime of Vesalius. The term is used by Vesalius in both editions of the

Latin   **Greek**   HEBREW   *ARABIC*

Fabrica (First edition, p. 524 ; Second edition, p. 644).   In neither does he mention Jacopo as its author.   The figure shows the prostate gland as though in front of the urethra.   The prostate had recently been described by Niccolo Massa of Venice in 1536 in an excellent account of the generative organs.   Vesalius could hardly have missed reading this.

TABULA II.

Description of the *vena cava* [34], *jecoraria* [35], **koilē**, HA-ORTI [36] *hanabub* [37], by which blood, nutriment of all the parts, is distributed throughout the entire body [37].

34) *Vena cava* = hollow vein.   Why hollow, since all veins are obviously hollow ?   The answer is that the Latin *cava* was a rendering of the Greek **koilē**.   This, while signifying primarily hollow, comes secondarily to mean soft, yielding, as hollow objects are liable to be.   Hence **koilia** = belly (whence our ' coeliac ' and ' coelom ').   Thus, **phleps koilēs** = belly-vein = *vena cava*.   In medieval anatomy **koilēs** becomes *chilis* and *vena cava* = *vena chilis*.

(35) *Jecoraria*, a barbarous adjective from *jecur, -oris* (Pliny, *jocur*), etymologically equivalent to Greek **hēpar**.

(36) The Hebrew letters spell HA-ORTI.   This substitution of the equivalent of aorta for *vena cava* confirms that in 1538 Vesalius was ignorant of the phonetic value of the Hebrew letters. They spell nothing like the next word, *hanabub*.   (See next note and also note 71.)

(37) *Hanabub* is a transliteration of the word in Hebrew characters at the head of Tab. III.   The word is the biblical HA-NABHUBH = the hollow [vessel], translating *AL-AJWAF* = **phleps koilē** = *vena cava*.   (See note 34.)   In the *Fabrica* (p. 376 *bis*, 4 lines from end) Vesalius repeats as a name of the vena cava *in Arabum autem interpretibus vocatur hanabub, vena ventrem habens*, ' by the translators of the Arabs it is called *hanabub*, that is, the vein having a belly '.   He must mean here the Hebrew translators, since no Latin translator uses this word.   Moreover, among Hebrew translators only Meathi and, following him, Lorci employ HA-NABHUBH.   The printed Hebrew Avicenna uses instead HE-ḤALUL. Therefore Vesalius in using this word *hanabub* was relying on advisers who were using manuscript sources.

The Liver, Source of the Veins [38].

(38) The liver, as source of the nutrient blood, which it distributes throughout the body through the veins, of which it is the fount, is one of the keys to Galenic physiology.

A. Veins behind the ears and to the temples.   (' Posterior auricular '.)
B. To the nostrils, forehead and upper jaw.
C. To the tongue, larynx, fauces and palate.   (' Common facial '.)
D. Internal jugulars, apople[c]tic, *profundae* [39].

I cannot depict here how the internal jugular and that [vein] which runs through the transverse processes of the cervical vertebrae (' Vertebral ') ramify in the brain and its membranes and ventricles.

(39) Jugular, *jugum* = yoke.   In classical writings *jugum* is used of the clavicle (Celsus) and hence of the hollows above the clavicles, the ' salt-cellars '.   Hence *jugulare*, to strangle.   The ' vein that runs in the hollow of the neck ' is called *vena jugularis* in medieval translations of Galen, rendering **sphagitēs phleps** or *guidez*.   (See note 40.)

By Mundinus (1316) the jugulars are called *Venae apoplecticae quia ex plenitudine earum frequenter sit apoplexia* ' because apoplexy often comes from their engorgement '. **Apoplēktos** = stricken down, disabled by a stroke.   The omission of the ' c ' in apoplectic here and in the next line, along with many similar misprints, indicates that Vesalius either saw no proofs or that he failed to read them carefully. The point is important in considering the conditions of publication. (See note 344 on colophon.) The term *profundae* is also applied to the jugulars by Mundinus, *quia sunt locatae in profundo juxta sive supra musculos et spondilia colli*, 'because they are placed deep down, near or over the muscles and vertebrae of the neck'.

E. External jugulars, *guidez*, which they also call *apople[c]ticae* [40].

(40) *Guidez* is a familiar medieval anatomical term which originates in Gerard of Cremona's Latin translation of Avicenna's *Canon*. It corresponds to *WADAJ*, pl. *WIDĀJ*. On *apoplecticae*, see previous note.

F. To the posterior muscles of the neck. ('Posterior external jugular'.)

G. Through the transverse processes of the vertebrae of the neck. ('Vertebral veins'.) They run into the spinal medulla and cerebrum.

K. To the convexity of the scapula and neighbouring parts. ('Transverse cervical'.)

L. *Humeraria* [41], outer side of arm, *cephalica, capitis* [42]. ('Cephalic'.)

(41) The first use of *humeraria*, translating **ōmolaia**, is by Guenther of Andernach in his translation of Galen's *De anat. admin.*, Paris, 1528, III, Ch. 5. The same term is used by him in his *Inst. anat.*, Basel, 1536, whence Vesalius took it. The corresponding term *AL-KATFĪ* (see note 42) was rendered *spatularia* by Gerard of Cremona.

(42) The history of the terms cephalic vein and basilic vein is a puzzle. Despite similarity to **kephalikē** and **basilikē**, the words did not come into Latin direct from Greek, nor are they known in any Greek author. Their first traceable use is in the Arabic translation of Galen's *De anat. admin.* Bk. X, by Ḥubaish b. al-Ḥasan and Ḥunain b. Isḥāq (ninth century) (Max Simon, 'Sieben Buecher Anatomie des Galen', Leipzig, 1906, vol. I, p. 58, and vol. II, p. 42). This suggests that the words were familiar in their time. Cephalic in the form *AL-QĪFĀL* is foreign to Arabic and its ultimate origin is unknown. It occurs first as gloss by Ḥunain on *AL-KATFĪ* (see note 41) = the vein belonging to the shoulder = **ōmolaia**. Avicenna distinguishes the upper part of the vein as *AL-KATFĪ* from the lower, *AL-QĪFĀL*. It cannot now be decided whether *AL-KATFĪ* = axillary vein or = the great branch that in the dog connects the cephalic with the external jugular. Basilic in its form *AL-BĀSĪLĪQ* is also not Arabic. Derivation from **basilikos** = royal is unintelligible, but no other has been suggested.

The Spanish Arab Abulcasis (d. 1106) in his *Chirurgia*, II, Ch. 95, and the North African Ibn al-Hashshā (13th cent.), in his glossary to Rhazes' *Almansor* (ed. G. S. Colin and H. P. J. Renaud, Rabat, 1941, p. 107) say that 'head vein' is a popular term. *Cephalic* and *basilic* were popularized in Latin by the translation of Haly Abbas by Stephen of Pisa (1127), by translations of Abulcasis and Avicenna by Gerard of Cremona (c. 1170), and by the so-called *Anatomia vivorum Galeni* (c. 1225), a Latin compendium of Arabic anatomy. (This last appears in the sixteenth century editions of the *Opera Galeni*.) The words came into more general medical use after 1300 with the translation of Avicenna's *Cantica* by Armengaud Blasius of Montpellier (d. 1314). This work normalized the elaborate rules of venesection.

M. To the front of the chest and breasts. ('Internal mammary vein'.)

N. To the upper muscles of the thorax.

O. *Axillaris*, inner side of arm : on the right called [vein] of the liver ; on the left it is called the [vein] of the spleen, *basilica* [43].

(43) *Axillaris*=basilica, see note 42. The term *axillaris* goes back ultimately to the **maschaliaia** of Paul of Aegina and was popularized by the translations of Gerard of Cremona. The basilic was opened on the right for diseases supposed to be situated in the liver and on the left for those of the spleen, a natural device with reference to the position of these organs.

| *Latin* | **Greek** | HEBREW | *ARABIC* |

This bifurcation of the *vena cava* within the thymus gland (*glandium*) [44] sometimes occurs a little lower, so that the axillary arises from a secondary branch, as does the *humeraria*.   Moreover, the veins of the chest which are also distributed to the mammae ('superficial epigastric') are then branches of the *axillares*. Further, the external jugulars may be double on either side of the neck [45].

(44) *Glandium* means here thymus.   *Fabrica*, 261*, lines 39–40, '*Quae vero in jugulo sub pectoris
    osse habetur . . . a Graecis* **thymos** *a Latinis communi voce glandium appellatur*' ('There
    is a structure in the neck under the breast-bone called in Greek *thymus* and in Latin in
    common parlance *glandium*').   This passage is practically repeated in *Fabrica*, p. 282,
    lines 18–19.   How the word became applied to the thymus is a puzzle, but it is so found
    in both the *De usu part.* of Galen, VII, 4 (Kühn, III, 424) and in the *De appell. part.* of his
    contemporary Rufus (168).   The term was adopted by Guenther and by Charles Estienne.
(45) The *vena cava* depicted by Vesalius is not that of man, but of an animal, presumably Macaca.
    In that animal the superior *vena cava* is much longer than in man, and the two *venae
    innominatae* are more symmetrical than in man.   Moreover, in Macaca up to a comparatively
    late stage the thymus is large and the innominates quite covered by it.

    P. Branch from *humeraria* to *media* [46].

(46) *Media*, translating Galen's **phleps mesē**.   Celsus renders it *vena ad medium*.

    Q. Branch from *axillaris* to *media*.
    R. To the elbow joint from *humeraria*.
    S. To the elbow joint from *axillaris*.
    T. *Media* [47], *communis* [48], *mediana* [49], *nigra* [50], *funis brachii* [51], *mater* [52].
        This begins at [the level of] the elbow joint and sometimes a little lower.

(47) For *media* see note 46.
(48) *Communis* is a popular medieval vein-name used by Mundinus.   Perhaps it translates *AL-UMM*
    (see note 52).
(49) Despite its Latin form, *mediana* as a vein-name is perhaps of Arabic origin.   The word *medianus*
    is unknown in classical Latin, except in Vitruvius, though it occurs in the vulgar tongue.
    In Ḥubaish's version of Galen's *De anat. admin.* (see passage in note 42) **phleps mesē** is
    translated *AL-AWSAṬ* = the middle [vein].   On the other hand, in the *Cantica* of Avicenna
    it is called *AL-MĀDHIYĀNI* = the two freely-flowing ones.   Armengaud (see note 42)
    transcribes this as *almadiam* in his Latin version.
(50) *Nigra* occurs first in Gerard of Cremona's translation of Avicenna, where it renders *AL-AKḤAL*
    = the blue-black or antimony-coloured.   This is the original Arabic name of the vein.
    Ḥubaish gives it as a gloss on *AL-AWSAṬ* (see note 49), the word adopted from the Greek.
    Nigra is used in the pseudo-Galenic *Anatomia vivorum* (c. 1225).
(51) *Funis brachii* was also introduced by Gerard of Cremona, translating Avicenna's *ḤABL
    AL-DHIRĀ'* = rope of the forearm.   To call a vein a rope was less strange to the
    Arabs than to us, for they also called the *humeraria ḤABL AL-'ĀTIQ* = rope of the
    shoulder.   These Arabic vein-names were probably originally butchers' terms.
(52) *Mater* we have not traced, but its origin must also be Arabic.   A hint as to the underlying idea
    is provided by the following passage in the dictionary of Ibn Sīda (1006–66) known as
    al-Mukhaṣṣaṣ : 'It has been said that *AL-AKḤAL* (see note 50) is the vital vein ('*IRQ
    AL-ḤAYĀT*), also called river of the body (*NAHR AL-BADAN*)' (I, 167).   A vein of
    that importance might well be called *AL-UMM* = the principal one.   Moreover, *AL-UMM*
    also = mother.   (See also note 48.)

    V. Various branches of veins in the terminal part of the limb.

The distribution of veins in the peripheral parts of the upper limb need not cause you too much anxiety since you will find scarce two cases alike in twenty, but the three chief branches are as represented.   Wherefore, and because of the remoteness and smallness of those veins, the Greeks did not open the veins in that part, save very rarely in chronic affections of the spleen. They then opened that which creeps between the *auricularis* and the *annularis* digit of the left hand, Y (53). This some call *Syelem* in both hands, while that which runs to the thumb, X, they call *Salvatella*. This [usage] is permissible, and we also find the former [*i.e.* Syelem] called by that name [*i.e.* Salvatella] (54).

(53) The medieval usage in naming the digits, beginning with the thumb, was *pollex, index, medius, annularis* or *medicus* or *cordis,* and *auricularis.*

(54) *Syelem* and *salvatella.* These strange-looking words and variants on them were used by medieval writers to describe the superficial veins on the ulnar side of the dorsum of the metacarpus, usually conspicuous in elderly subjects. Vesalius, like other early anatomists, has difficulty in distinguishing between these veins. There are two good reasons for this confusion. Firstly, they are so irregular that no rule can be laid down as to their distribution. Secondly, the two words, though they look different, are in origin identical and are frequently confused, as by Vesalius himself.

    Syelem is found, with many variants, as *seile, sceyle, sedem, alaseilem,* etc. Its ultimate source is the Arabic text of the *Canon* of Avicenna, *AL-USAILIM* or *AL-USLĪM,* a word of unknown origin. It is transcribed in Gerard of Cremona's Latin version as *sceilen.* This, through *alaseilem,* becomes *salaseilem,* as in the *Conciliator* of Peter of Abano (c. 1290). It had already passed into *salacella* in Albertus Magnus (c. 1270). Thence its passage into *salvatella* is easy. The letters *c* and *t* are interchangeable in most medieval scripts, and often in pronunciation, e.g. *natio=nacio,* etc.

A. To the upper four or sometimes three ribs. (' Highest intercostal '.)

B. To the lower eight ribs. By the Greeks called **azygos** (55), that is, without fellow. You may take it that this arises from the right side of the *vena cava.*

(55) **Azygos** is applied to the vein in Galen's commentary on Hippocrates, *De victus ratione in morbis acutis* (Kühn, XV, 529. See note 56). The word **azygos** is not used by Vassaeus in his Latin translation of this work of Galen, Paris, 1531, or in Gadaldino's revision of Vassaeus in the ' first ' Giunta Galen, Venice, 1541, nor does Nicholas Massa use the word in his *Anatomiae liber introductorius,* Venice, 1536. It was used by Guenther in the *Inst. anat.,* Basel, 1536, whence, doubtless, Vesalius took it. Thus Guenther must be credited with introducing the term **azygos** into anatomical terminology.

B. This unpaired vein [the **azygos**], which is described as nourishing eight lower ribs, we have never seen arise below the right auricle of the heart [in man], though in dogs and monkeys [it arises] a little above it (56). Wherefore it is better for pain in the side tending downward to employ venesection rather than purgative medicines (57). And as for the view of Hippocrates, I hold Galen to have spoken unclearly about this vein in the second book, *De victus ratione in morbis acutis.* Moreover, having regard to the place of origin of that vein and in view of the continuity and direction of its fibres or texture, it seems reasonable always to open the inner vein of the right elbow for pain in the side at the level of the third or fourth rib or lower. Now thoracic pain affects chiefly the middle region, therefore venesection should be considered rather for the right [than the left] side, a point which I could wish anatomists would consider more worthy of discussion (58).

Latin	**Greek**	HEBREW	*ARABIC*

(56) The statement as to the difference in man and in animals in level of junction of **azygos and**
vena cava is correct. We have verified this in Macaca. It is not, however, an observation
original to Vesalius, for it had been made by Galen. Morover, the words of Vesalius are an
almost word-for-word translation of a passage in which Galen shows himself quite conscious
of the differences between simian and human anatomy. (See Galen, *De victus ratione in
morbis acutis*, II, Ch. 10 ; Kühn, XV, 529.)

(57) Since the **azygos** supplies the lower thorax, evacuation is held to be best from the most convenient vein near the **azygos**. On the other hand, since the abdomen is supplied by the
portal, from which bleeding is impracticable, evacuation for that region is best by purgatives.

(58) In venesection for pain in the chest, that vein should be opened which, being convenient for
the operation, is also nearest to the source of the vein supplying the painful area. For
pain below the third rib on either side this would be the vein at the bend of the right elbow,
as nearest to the ' root ' of the **azygos** vein. For pain above the third rib, in view of the
separate origin of the higher intercostal veins, the side chosen for venesection should be
according to the site of the pain. The passage referred to is *De acut. morb. victu comm.*
II, Ch. X (Kühn, XV, 533). Galen's offences are to have described four rather than three
costal vessels as connected with the *vena cava* and to have failed to stress the right elbow
for venesection for the lower thorax. (See also note 294, on Vesalius' attitude to Galen.)
The suggestion of Vesalius that anatomists should discuss the matter more has to be read
in the light of the fact that the physicians, who were not anatomists, had discussed it
ad nauseam. This futile discussion occupied a large part of medical activities in the middle
half of the sixteenth century. The book on the subject by Vesalius is the *Epistola docens
venam . . . secandam*, Basel, 1539. This book places the discussion on an anatomical level,
though a false one. (See notes 19 and 32.)

C. The part of the [*vena*] *cava* that is continuous with the right cavity of
the heart.

D. *Coronal* vein called **stephaniaia,** which is sometimes double, just as seems
to be the case with the *coronal* arteries [59]. (' Coronary sinus and
tributaries '.)

(59) The history of the word *coronal* or *coronary* as applied to the vessels of the heart is obscure.
The passage above reflects Guenther's *Inst. anat.* Basel, 1536, p. 80, *Vena coronalis.
Secunda propago ex ramo illo* (*i.e.* of the *vena cava*) *in cor insigni antequam id ingreditur,
qui superficiem cordis coronae modo cingens, unde* **stephaniaia** *dicitur*, ' The second twig
of that great branch (of the *vena cava*) before it enters the heart, which embraces the surface of
the heart like a *corona*, whence it is called **stephaniaia** '. But the *coronal* vessels are not
at all like a crown or wreath. Guenther's passage is, in effect, a quotation from Galen, *De
usu part.* VI, Ch. 14 (Kühn, III, 477), whi h says ' this vein encircles (**peristephanousa**) the
heart.' The association with a crown is probably Guenther's, who thus introduced it into
ana omical nomenclature.

E. Veins of the transverse septum, which have sometimes been found triple.
(' Inferior phrenic veins '.)

The first division of the *vena cava* is in the very substance of the liver and not
outside it, if we are to call it a division at all. This I would merely state, for I
would avoid the circumlocutions of the anatomists [60].

(60) That is, the junction of the hepatic veins with the *vena cava*. But who are these circumlocutory
anatomists ? Estienne's figure and account of a *vena chilis* emitting an upper and lower
branch was in print by 1539 and probably prepared before Vesalius left Paris (Fig. 9).
There is, however, a strong presumption that the two men were not in contact. But
Nicholas Massa (*c.* 1500–69) of Venice was an obvious rival to Vesalius. His *Anatomiae liber
introductorius*, Venice, 1536, must have been known to Vesalius, though we cannot find that
he anywhere mentions it. This book contains two long chapters on the vena chilis. To it,
perhaps, Vesalius here primarily refers. In the *Fabrica*, 1543, p. 277, corresponding to the
1545 edition, p. 458, is a detailed discussion of the hepatic course and disposition of the
vena cava.

G. To muscles of the spine and parts near the spleen.

H. To fatty tissue of the kidneys. ('Supra-renal veins'.)

I. Passing down (*deferentes*) serous blood to the kidneys, they are called *emulgentes* [61]. ('Renal veins'.)

[H, I.] To make apparent the course of the emulgent [61] vein within the kidney we have depicted the kidney itself only on one side.

(61) It was supposed by the medieval and renaissance anatomists that the blood emerging from the liver shed its various superfluous elements, 'impurities' or 'fumes' by different routes, each into its special organ. Thus the black bile was supposed to be evacuated from the spleen to the stomach. (See note 30.) Similarly the *emulgentes* = 'milking veins' (the root—*mulg* = English 'milk') drain off the watery element of the blood into the urine, taking with it some of the choler or yellow bile. Hence the colour of the urine. The term *venae emulgentes* is known first from Mundinus (1318). On physiological theory Mundinus naturally recognised only veins as emulgent, not arteries. To call an artery 'emulgent' would deprive the word 'emulgent' of its meaning in medieval physiology. This, however, Vesalius does. (See note 87.)

K. Left [*vena*] *seminalis*, which sometimes draws a branch from the *vena cava* which unites with it [62]. ('Spermatic vein'.)

(62) This variation is figured in the *Fabrica*, 1543, p. 209, and second edition, p. 474.

L. Right *vena seminalis*. ('Spermatic vein'.)

M. To each of the lumbar vertebrae. ('Lumbar veins'.)

N. To the lumbar muscles and to the transverse and oblique [muscles] of the abdomen. ('Ilio-lumbar veins'.)

O. To foramina of the *os sacrum* [63]. ('Lateral sacral veins'.)

(63) On the term *os sacrum* see note 324.

[O] Here the continuity of the vessels of the two limbs can easily be seen; it is otherwise in the [thymus] gland [64].

(64) Vesalius thinks of the *Vena cava* as a single trunk running the length of the body and bifurcating above much in the way that it does below. This is much more nearly true in Macaca, with its two equal and almost symmetrical innominate veins than in man. Moreover, in Macaca the openings of the superior and inferior *venae cavae* into the right atrium are so close together that they give the impression that a single main vein runs the whole length of the body (Fig. 44). This gives more point to the comparison by Vesalius of the bifurcation in the thymus to that in the pelvis. See, however, p. lii.

P. To the *intestinum rectum* [65] and surrounding parts. ('Haemorrhoidal veins'.)

(65) *Rectum* is the term used by Celsus. It is similarly employed by Gerard of Cremona in his translation of Avicenna's *Canon* (c. 1170) and adopted by Mundinus (1318) as an alternative to the more usual medieval Latin term *longanon*. It is also used by the later translators in rendering **apeuthysmenon enteron** (= stretched or straightened intestine) of Galen, Dioscorides and Oribasius. It seems a strange choice, for the human rectum is far from straight. But it must be remembered that in the animals specially dissected by Galen, Rhesus monkey, dog and pig, this structure is straight.

Latin	**Greek**	HEBREW	*ARABIC*

Q. To bladder and uterus.   (' Vesical veins '.)

R. To penis or to the *collum* and *fundus vulvae* [66].   (' Internal pudendal veins '.)

(66) The medieval view of the structure of the female genitalia is a complex theme. Vesalius develops it somewhat along the traditional line of an homology of the male and female organs.   This may be gathered from his figures on Tab. I, with which the following scheme may be compared :—

Vulva								*Urinary meatus*
Vagina	..	..	..	..	..	..	..	*Penis*
Collum	..	..	..	..	..	..	..	*Neck of bladder*
Cornua	..	..	..	..	..	..	..	*Prostate*
Uterus	..	..	..	..	..	..	..	*Bladder* !
Testis mulierum [Ovary]	..	..	..	..	..	..		*Testis*
[Fallopian tube]	..	..	..	..	..	..		*Vas deferens*
[Plexus of ovarian vessels]	..	..	..	..	..			*Pampinniform plexus*

The relation of these various parts in medieval and renaissance anatomy we have traced more fully elsewhere.   Charles Singer, the *Fasciculo di Medicina, Venice*, 1493. Florence, 1925, 2 vols. Vol. I, p. 255 ff.

S. To the pubes and to the transverse muscles of the abdomen.   A large part of this [vein], having gone to the *recti* [67] [*abdominis*] muscles, goes on to join the veins of the chest.   (' Inferior epigastric vein '.)

(67) The abdominal muscles were a subject of constant interest and are described in most of the medieval anatomical texts.   Mundinus (1318) calls the *recti* the *longitudinales*.   Berengario da Carpi (1523) and Massa (1536) call them *longi*.   The old translation of Galen's *De usu part.* by Nicholas of Reggio (*c.* 1320) uses the term *recti*, and is followed by Guenther in his translation of Galen's *De anat. admin.*, whence Vesalius doubtless derived it.

S. How these [epigastric] veins communicate with the veins of the chest is clear enough, but cannot be represented in the plate.

T. To the outer muscles of the thigh.   (' Deep iliac circumflex '.)

V. Through the limb to the lower part and as far as the foot.   (' Femoral veins '.)

X. To hip joint and to the outer part of the thigh.

Y. It extends under the skin through the inner part of the limb as far as the foot.   (' Great saphenous vein '.)

A. These two branches, arising from the great vein to the middle of the limb, form the *vena poplitis*.

B. From the *vena poplitis* to the skin on the outer side of the limb.

C. Division concealed in the popliteal space.

C. This bifurcation in the popliteal space sometimes seems to arise as three fairly large veins.

D. To the skin of the calf in which, as in the po[p]liteal veins, varices are liable to arise.

E. Wanders over the external malleolus [68] and is diffused on the outer part of the foot and called [*vena*] *schiatica* (' small saphenous vein ') because used in treatment of hip-joint disease.

(68) On the senses and use of the word *malleolus*, see note 264.

F. Wanders over the internal malleolus and on to the inner side of the foot and called *vena matricis* because used for discharging uterine humours (*mala*).   *Saphena* [69].

(69) *Saphena.* Despite its Greek homophone **saphēnēs** = clear, obvious, in the intellectual sense, this word is of Arabic origin. It appeared first in Gerard of Cremona's translation of the *Canon* of Avicenna (c. 1170). In the original the term was *AL-ṢĀFIN*, derived from *ṢAFANA*=to stand with the heels raised above the ground, applied especially to horses. It is not clear what the vein has to do with this posture. The same name was also applied to some branch of a vein in the thigh (Ibn Sida, Mukhaṣṣaṣ, II, 49), as well as to one in the leg (*ib.* 53). Ibn al-Hashshā' (13th century) says : '*AL-ṢĀFIN* is a vein which stretches along the thigh and into the leg on its inner side towards its back. It is bled near the heel on the side of the big toe ". That is the long saphenous vein. A similar but less clear description is given by Haly Abbas. The internal saphenous vein was very frequently employed for venesection and was especially recommended for uterine trouble. Hence its common medieval synonym, mentioned by Vesalius, *vena matricis.*

Just as there is a manifold spread of the veins at the extremity of the upper limb, so in the lower. Wherefore, as we read, the Greeks opened the veins either at the malleolus or behind the knee and did not open them at the extremity of the foot, as others have done, cutting there ineffectually without evoking blood-flow. Moreover [the Greeks] avoided the veins from the [*vena*] *poplitis*, as it cannot be discerned to which toes they go because of thickness of the skin [70].

Some reckon the major branches of the *vena cava* as 168.

(70) The whole sixteenth century doctrine of venesection and the portentous and disputatious literature that rose around it originates in Galen's *De curandi ratione per venae sectionem* (Kühn, XI, 250). It was translated into Latin during the fourteenth century, but then could have exercised very little influence, since only two copies are known. It appeared very early in print, however, for it forms a member of the first edition of Galen's collected works, in the Latin version of Diomede Bonardo of Brescia, Venice, 1490. The standard translation of the sixteenth century was that of Theodorico Sandano, which is reprinted in the Giunta editions.

TABULA III.

Arteria magna, **aortē** [71], HA-NABHUBH [72], *haorti* [73], arising from the left cavity of the heart and carrying the vital spirit [74] to the whole body, regulating the natural heat by contraction and dilatation [75].

(71) The passage of the word Aorta to its modern anatomical meaning is intricate and significant for the history of nomenclature. It is connected with a root which carries the meaning of 'raising on high' (**aeirein**). **Aōr** appears in Homer, where it means a sword-belt or strap, and especially a cord from which something is suspended for exhibition (*Odyssey,* XI, 609 ; XIII, 438). Hence **Aortē** passed into a butcher's term for the stem of the ' pluck ', that is of the main collection of viscera—heart, lungs, etc.—pulled from the carcase of a beast. Models of such plucks are represented in several ancient votives. In the *De Corde* (Ch. X), a member of the Hippocratic Collection of the early fourth century B.C., **aortē** is used in this way in the plural or collective for the complex of great vessels attached to the root of the heart. In Aristotle's *Historia animalium* of about half a century later, **aortē** in the singular has come to mean the great vessel arising from the left ventricle (496 *a*, 28, and elsewhere). Galen called this vessel **artēria megalē** or **megistē**, hence the *arteria magna* of this plate. He also calls it **artēria orthē** = straight artery, but various medieval contortions such as *haorti, adorti, ahorti,* etc., probably go back not to this, but to the Hebrew and Arabic forms, *cf* note 73.

| *Latin* | **Greek** | HEBREW | *ARABIC* |

(72) The equivalent of *haorti* in Hebrew letters has been by error interchanged with its fellow in the heading of the previous plate (see note 36), and absurdly transliterated *hanabub*, while in this plate the Hebrew letters do, in fact, spell HA-NABHUBH.

(73) *Haorti*, printed as though transcribing the previous Hebrew word, is not the form in which *aorta* is normally found in Hebrew writings. There it is ORITI, the *orithi* of Gerard of Cremona's translation of the *Canon* of Avicenna or, as vocalized in the Oxford MS of Graciazo's version, AWRITAY. This latter form may underlie the *aurithia* of the Latin Haly Abbas of Stephen of Pisa (1120). Note that both Gerard and Stephen transcribed the word in a form current only among Jews, although it was the Arabic text that they had before them. This indicates that their work was done with Jewish help. The form HA-ORTI, spelt as in our text, not HA-ORITI, with the definitive article HA- added, occurs once in the Oxford MS of Lorci's translation. It may represent the pronunciation used by Italian Jewish doctors.

(74) The 'vital spirit' is the second of the Galenic spirits. It was supposedly formed in the left ventricle from blood that came through the septum, bearing natural spirit which mixed in the ventricle with 'pneuma' from the outer air. This pneuma was supposed to be brought to the left ventricle by the 'venous artery' (that is the pulmonary vein).

(75) The expansion and contraction of the arteries was well known to the ancient and medieval writers. It was believed to be an active process controlling the bodily heat. When the pulse quickened it thereby caused temperature to rise.

The heart, the source of the vital faculty and the origin of the arteries [76]. [Written on the heart itself.]

(76) See note 74. In Galen's physiology the veins originate not in the heart but in the liver. Thus the right ventricle and its supposed branch, the 'arterial vein' (= pulmonary artery), are really branches of the *vena cava*. On the other hand, the venous artery (= pulmonary vein), like the aorta, is held to arise from the left ventricle.

A. *Plexus choriformis* in the anterior ventricles of the brain formed from arteries and veins [77]. ('Anterior chorioid plexus'.)

(77) The *plexus choriformis* is the chorioid plexus of the lateral ventricles of the brain. Vesalius here normalises the term in anatomical nomenclature He has taken it from Guenther who, writing it always in Greek script, uses it in his translation of Galen's *De anatomicis administrationibus*, Paris, 1531, IX, Ch. 3 (corresponding to Kühn, II, 719) and again in his *Institutiones anatomicae*, 1536. Galen uses the term **chorioeidē plegmata** only this once, relating that it was given by the followers of Herophilus because it looked like the embryonic membrane known as **chorion**, The original meaning of **chorion** as employed by Aristotle is the membrane inside the shell of an egg.

The term **chorioeidēs** is often applied by Galen to the *pia mater* and also to the chorioid tunic of the eye, supposed to be an outgrowth of the *pia*. In this sense the word enters a Latin text, though still in Greek script, in Guenther's translation of the pseudo-Galenic *Introductio seu medicus*, Paris, 1528.

B. *Plexus reticularis* at the base of the brain. The *rete mirabile* in which the Vital is elaborated into Animal Spirit [78].

(78) A remarkable plexus of blood-vessels is formed from branches of the carotid at the base of the brain in carnivors, and is even more conspicuous in ungulates. This structure greatly impressed Galen. He describes it as a **diktyoides plegma** = net-like wreath (**diktyon** = fishing-net), and devotes a whole chapter to it in his *De usu partium*, IX, Ch. 4 (Kühn, III, 696–700), and speaks of it as **thauma megiston** = *maximum miraculum*. The *De usu partium* was made available in Latin by Nicholas of Reggio about 1320, but an abbreviation of it, containing an account of the *rete mirabile*=wonderful net, was available from about 1250 under the title *De juvamentis membrorum*. This was much more popular than the longer work, and was familiar to anatomists from Mundinus (died 1326) onwards.

The conception of a *rete mirabile* was essential to Galenic physiology. Blood from the heart bearing the Vital Spirits was brought to the rete, there to be charged with Animal Spirits. It is these last that pass into the nerves and induce muscular movement.

It will be noted that Vesalius does not here doubt the existence of a *rete mirabile* in man. But Berengario da Carpi in his *Isagogae breves . . . in Anatomiam*, Bologna, 1523, having described the *rete* and its action, wrote : ' Nevertheless I have never seen this *rete*, nor do I believe that Nature works with many instruments that which she can achieve with few [' Ockham's razor '] . . . and therein I take experience for my charioteer '. Johannes Eichmann (Dryander) in his *Anatomiae*, Marburg, 1537, after a lengthy and illustrated account of the brain, ends with the phrase : 'At the base of the brain, that is on the bone underlying it, contiguous to the plate, thou wilt find the *plexus reticularis* formed from arteries entering the head '. His figures of the brain are poor, but drawn from nature and, unlike Vesalius, he does not represent the *rete*. Charles Estienne, writing at the same time as Vesalius, says that the *rete mirabile* can only be seen in bodies of those very recently dead or who have died suddenly. Nicholas Massa (1536) writes hotly of doubters : ' Some dare to say that this *rete* is a figment of Galen . . ., but I have often seen it and demonstrated it ' (Chapter 39).

C. Arteries behind the ears and to temples and face.

D. To tongue, larynx and fauces.

E. *Arteriae* **karotides,** that is the sleep producers, *apople[c]ticae* [79], *subeticae* [80], HA-NIRDAMIM, *hanirdamim* [81].

(79) **Karotides** is the name repeatedly applied to these vessels by Galen. Celsus uses the word once. Rufus of Ephesus, who was a century before Galen, but whose works were inaccessible to Vesalius (*editio princeps*, Paris, 1554) derives the word from **karoein,** to send to sleep, render unconscious, stupefy. There was from an early date a view that pressure on these arteries causes unconsciousness and a condition described as ' apoplexy '. This is expressly denied by Galen on the basis of experiment (*De usu resp.* Ch. 5 ; Kühn, IV, 502–3). The word ' carotid ' was well known to medieval anatomical writers, whom it probably reached through the thirteenth century Latin Galenic compendium *De·juvamentis membrorum.* See note 78. Mundinus says of the carotids that ' they are named apoplectic because from their repletion apoplexy frequently ensues. They are also called *venae somni*, because from obstruction in the *rete mirabile* sleep is induced '.

(80) The same idea is expressed in the Latino-Arabic *subeticae*, that is *AL-SUBĀTIYYĀNI =* the veins causing torpor, from *SUBĀT =* torpor. *Subeticae* entered Latin medicine with Gerard of Cremona's translation of the *Canon* of Avicenna. The conjunction of the words *carotidae, soporariae, apoplecticae, subeticae* used by Vesalius had been made by Andrea Alpago of Belluno in the 'first' Giunta Avicenna, Venice, 1527. From that book (I, 1 ; V, 4 ; 4) Vesalius therefore probably took the phrase.

(81) HA-NIRDAMIM, the stupefied [arteries], a none too apt translation of *AL-SUBĀTIYYANI*, is used by the Avicenna translators Meathi and Lorci. Graciano and the printed Hebrew Avicenna merely transcribe the Arabic word. The Niph al verbal form of the root RADAM = to sleep, is employed in the sense of ' to be quite unconscious ' in Psalm lxxvi. 7, and Daniel, viii. 18 (RV both times=in a deep sleep).

F. Running through the transverse processes of the vertebrae of the neck as far as the brain. (' Vertebral arteries '.)

G. To the *os pectoris* and to the mammae which communicate with those which are in the *recti* [*abdominis*] muscles [82]. (' Internal mammary arteries '.)

(82) See note 67.

Latin	**Greek**	HEBREW	*ARABIC*

H. To the humeral muscles and convexity of the scapulae. (' Transverse
     scapular arteries '.)

I. To the superior costal muscles and to the mammae.

K. Runs under the axillary vein into the arm. (' Profunda brachii '.)

L. To the joint of the elbow, one on either side. (' Ulnar collaterals '.)

M. In the inner part of the hand (' Ulnar ') and a twig to the outer side of
     the thumb (' Superficial volar '.)

N. To the upper costal spaces of the chest. (' Superior intercostals '.)

O. The great division, of which the major branch (' Aortic arch ') is dis-
     tributed to the lower part of the body and from which branches are
     immediately given off to each of the ribs [83].

(83) It is very significant that the aorta and branches here depicted are certainly not human. They
       are those of an ape. The ape in the frontispiece of the *Fabrica* is a *Macaca mulatta* (Rhesus
       monkey) which, though closely allied to, is not identical with *Macaca inua* (Barbary ape)
       which Galen dissected. The inner anatomy of the two species would certainly not be
       distinguishable.
           Fig. **44** represents the heart and aorta of a young female *Macaca mulatta* of the natural
       size. The heart of the Macaca monkey presents an elongated oval outline, the long axis
       of which is much nearer the vertical than is the long axis of the human heart. The right
       ventricle lies clasping the left ventricle, pursuing a spiral course upward and forward
       to the very prominent pulmonary artery. The right ventricle has an apex that projects
       slightly from the contour of the heart. The left coronary artery is much larger and more
       conspicuous than the right, and soon after leaving the left anterior sinus it divides clearly
       into two main branches.
           The aorta of Macaca passes away from the heart by a sharp curve down into the
       mediastinum in an almost direct antero-posterior plane. From its arch only two branches
       arise. The first branch, the *truncus communis*, is very large and arises from the arch just
       outside the attachment of the pericardium. It has no counterpart in man. From it arise
       the left common carotid and the innominate. The latter passes back to the neighbourhood
       of the right first rib, where it divides into the right common carotid and right subclavian.
       The second and smaller branch of the aorta, the left subclavian artery, comes off from the
       crest of the arch and proceeds almost upward for a considerable distance before emerging
       from the thorax. (This account, which corresponds to our findings, is slightly modified
       from that of P. Lineback in C. G. Hartman and W. L. Strauss, ' The Anatomy of the Rhesus
       Monkey ', London, 1933.
           It is remarkable that Vesalius persists in his erroneous account, drawn from the anatomy
       of the monkey, of the branches of the aortic arch in the *Fabrica* (see diagram in 1543 edition,
       p. 295 ; second edition, p. 483). Nevertheless, even in the *Tabulae Sex* he shows awareness
       that there is some defect in his description. See the following two paragraphs.

We have sometimes seen the left carotid given off from that [artery] which
goes to the left arm so that both divisions [of the arteries] of the chest are given
off in the same way as that which goes to the right limb [*i.e.*, there are two
innominate arteries].

The unequal division of the *magna arteria* (*i.e.*, ' Ascending Aorta ') is sometimes very near the heart and sometimes somewhat further from the heart as
here portrayed.

The origin of the coronal arteries cannot be depicted in this plate for they
are hidden beneath the valves (*membranulae*) which prevent the [Vital] Spirit
from returning from the *magna arteria* to the heart [84].

(84) The passage shows how fully the nature and action of the sigmoid valves in the great arteries were appreciated by the older anatomists. This in no way interferes with the acceptance of Galenic physiological theory, as Vesalius himself indicates. It is noteworthy that the far less perfect system of venous valves was the starting point of Harvey for his destruction of that theory.

P. *Vena cava* opening into the right sinus of the heart.

Q. Venous artery conveying air from the lungs into the left sinus. (' Pulmonary vein '.)

R. Arterial vein conveying the blood into the lungs from the right sinus. (' Pulmonary artery '.) (85).

(85) In ancient nomenclature the pulmonary artery is called a vein as being a diverticulum of the right ventricle, which is itself but a branch of the great venous stem emerging from the liver, the *principium venarum* (Tab II) via the ' auricula '. Similarly the pulmonary vein is called an artery as ' emerging ' from the ' appendix ' or ' auricula ' of the left ventricle.

S. Rather large arteries of the transverse septum. (' Phrenics '.)

T. Into the *simum* of the spleen and in size proportional to that viscus. (' Splenic artery '.) (86).

(86) The size of the splenic vessel was of great importance in the old physiology, since the spleen was the organ of evacuation of the ' black bile '. On *simum*, see note 21.

V. To the fissure of the liver and to the gall-bladder. (' Hepatic '.)

X. To the stomach and omentum. (' Gastro-duodenal '.)

Y. In the upper part of the mesentery. (' Superior mesenteric '.)

The arteries which pass into the liver, spleen, stomach, *omentum* and mesentery [T, V, X] sometimes send out two roots as here, sometimes three, sometimes one [='Coeliac axis '], though rarely so in man. We have nearly always found them transversely arranged as in the figure.

A. To the kidneys, called *emulgentes*, smaller than the corresponding veins (87). (' Renal arteries '.)

(87) On the misuse of the term *emulgentes* for arteries, see note 61.

B. *Arteriae seminales*, one on each side (' Internal spermatic arteries '). As explained above, I have found the *arteriae seminales* always to be given off from the trunk of the aorta, thus differing from the *venae seminales*. The left one I have occasionally found to be actually wanting (88).

(88) See note 33.

C. Distributed via the mesentery throughout the intestines. (' Inferior mesenteric arteries '.)

D. To the lumbar vertebrae, to the transverse and oblique muscles of the abdomen. (' Lumbar arteries '.)

Latin	**Greek**	HEBREW	*ARABIC*

F. To the foramina of the *os sacrum* [89].   Some are wrongly accustomed
to show these as *venae haemorrhoidales* [90].   (' Middle sacral, etc.')

(89) On the term *os sacrum*, see note 323.
(90) On the source of the haemorrhoidal vessels, see note 19.

G. To the bladder, in men to the penis, in women to the *fundus uteri* and
collum vulvae [91].   (' Vesical arteries '.)

(91) On these terms and their analogies, see note 66.

H. Arteries through which the spirit is conveyed to the foetus in the uterus,
and they are sometimes clearly branching from (*implantatae*) the greater
vessels.   (' Obliterated umbilical ' [92].)

(92) There was considerable medieval discussion as to how and at what period after conception
the three spirits enter the foetus, that is to say, when and in what sense it became alive.
The matter had theological bearings, since on it depended the age at which it was
necessary to baptize an untimely birth.

I. To the *recti abdominis* muscles [93].   They are united with arteries of
the chest and through them there is communication between **mammae**
and uterus [94].   (' Internal epigastric arteries '.)

(93) On the name *rectus*, see note 67.
(94) The conventional medieval explanation of the ' sympathetic ' relation of these organs.

K. To the hip-joint and the outer region of the thigh.   (' Profunda femoris '.)
L. Bifurcation in the upper part of the popliteal space.   (' Division into
posterior tibial and peroneal arteries '.)
    We have seen this artery to remain undivided as far as the middle
of the leg.
M. Concealed in the inner part of the foot.   (' Posterior tibial arteries '.)
N. Distributed to the outer part of limb (called *profundae*) (' Peroneal ') and
from them very small branches pass to the upper side of the foot
(' Perforating of peroneal '), but they send a conspicuous twig to the
outer side of the great toe (' Arcuate ').
    This distribution of arteries in foot and malleolus has been found to
vary as [have the corresponding vessels] in the hand.   We have here
set down what we have most often seen.
    One hundred and forty-seven branches of the *magna arteria* seem
worthy of record.

TABULA IV.

The bones of the human body represented from in front.

The foramina visible in the figures in these three plates are : [one] in the
bones of the temples [95] [namely] the auditory meatus : one behind the mamillary
('Mastoid ') [96] process through which the internal jugular [97] plunges into the
brain : four in the face around the socket of the eye, the first to the forehead
(' Supra-orbital '), the second to the nose (' Lachrymal '), the third to the superior
maxilla (' Infra-orbital '), the fourth to the temporal muscle (' Inferior orbital

fissure') : two in the inferior maxilla (' Mental ' and ' Mandibular '), through each (*singula* for *singulatim*) of which emerges a branch of the third pair of nerves [*i.e.*, of the ' Trigeminal '].

(95) *Tempora* for the temple or bones of the temple is classical (Virgil, Lucretius, Pliny, Celsus, etc.), and passed early into most European languages with a tendency to assimilate an *l*, perhaps from an association of ideas with *templum* or perhaps from *os temporalis*. It is the same word as *tempus, tempora*, time, and appears to render **ta kairia** = the spot, the right spot, the fatal spot, the temples, but also carrying the sense of time. Vesalius thus follows the ordinary usage.

(96) The name given to this process by Galen is **mastoeidēs** = breast-like, which the medieval translators render *mamillaris*. Vesalius was perhaps the first to use the term ' mastoid ' in a Latin work (*Fabrica*, 1543, p. 31, line 36 ; p. 167, line 1. He spells the word in Greek script in both editions).

(97) For the word *jugular*, see note 106.

(98) The nomenclature of the cranial nerves employed by Vesalius is that of Galen. It may be set forth thus :—

MODERN USAGE.	GALEN.
I. Olfactory.	Not regarded as separate nerves.
II. Optic.	' The soft nerves of the eyes '.
III. Oculomotor.	' The nerves moving both eyes '.
IV. Trochlear.	Not described.
V. Trigeminal.	{ ' Third pair of nerves '. { ' Fourth pair of nerves '.
VI. Abducent.	United with II.
VII. Facial. } VIII. Auditory. }	' Fifth pair of nerves '.
IX. Glossopharyngeal. } X. Vagi. } XI. Spinal accessory. }	' Sixth pair of nerves '.
XII. Hypoglossal.	' Seventh pair of nerves '.

Odontes, *dentes*, SHINNAYIM *scinaim* (99) are at most thirty-two in number ; eight **tomeis** (100), *incisorii*, MĚḤATTĚKHIM *mechathchim* (101) ; four **kynodontes**, *canini*, KALBIYYIM (102) *calbiim* ; and twenty **myletai**, *molares*, *maxillares*, TOḤĂNIM (103) *thochnim*, which all penetrate their alveoli with unequal roots.

(99) SHINNAYIM (singular SHEN), ' teeth ', biblical.

(100) **Tomeis**, a Galenic term used in the *De ossibus* (Kühn, II, 754), one of the most elementary and most frequently read of Galen's works.

(101) MĚḤATTĚKHIM, ' incisors ', from ḤATTEKH (Mishnaic) = to carve up, cut to pieces = QAṬṬĀʻA, used by Haly Abbas and formed after Greek **tomeis** = cutters, but not used by Avicenna. (See note 100.)

(102) KALBIYYIM, canines, from KELEBH = dog, seems formed after **kynodontes**, *canini*, since Arabic uses *NĀB*, plural *ANYĀB*, a word not etymologically connected with the dog. Meathi and Lorci simply transcribe the Arabic word. Graciano says that they are called in Italian *canini*, but then goes on to speak of them as QALBIYYOTH. Perhaps he connected the word with *QALB* = heart. The word occurs in the MS Hebrew translation from the Latin of the *Almansor* of Rhazes (Bodleian MS Neubauer 2090), where it translates *canini* of the Latin text. The use of the word QALBIYYOTH by Graciano shows that it was current among Jewish physicians as early as the thirteenth century. In the printed Hebrew text

Latin   **Greek**   HEBREW   *ARABIC*

of the *Canon* of Avicenna (Naples, 1491), both terms are replaced by MĔTHALLĔ'OTH, which in the Bible refers to the (canine ?) teeth of the lion. (Meathi and Lorci use MĔTHALLE'OTH or its equivalent MALTA'OTH for the molars.) The *Fabrica* (1543, p. 167) gives MĔTHALLĔ'OTH, but adds KALBIYYIM, doubtless because this had become the current Hebrew word.

(103) ṬOHĂNIM = grinders, from ṬAHON = to grind, probably a popular rendering of **myletai**, *molares*. But it may also go back to Graciano, who thus renders *ADRAS*, perhaps taking his clue from the remark in the Arabic Avicenna that the *ADRĀS* are *LI-L-ṬAHN* = for grinding. It occurs in the MS Hebrew translation from the Latin of the *Almansor* of Rhazes (Bodleian MS Neubauer 2090), where it translates *molares*. Meathi and Lorci use MĔTHALLĔ'OTH (see preceding paragraph).

The *Tabulae* do not mention the wisdom teeth. These are given in the *Fabrica* with their Arabic name *NAWĀJIDH*, taken from the printed Hebrew Avicenna.

Note.—All these Hebrew words for teeth are adjectives in the masculine plural. SHEN is feminine in the Bible, but is here treated as masculine in accordance with medieval usage. The printed Hebrew Avicenna (Naples, 1491) uses instead of ṬOHĂNIM the grammatically more correct ṬOHĂNOTH, and this form appears in the *Fabrica*.

B. *Claviculae* [104], **kleides** [105], *claves, jugula* [106], TARQUWA *tharkuha* [107], *furculae* [108]. Each bone suggests the letter S. The figure is not symmetrical [109].

(104) *Clavicula*, diminutive of *clavis* = key, seems a poor description of the bone and is not classical. **Kleis** = primarily a bar or bolt, goes right back to the Iliad in the sense of collar-bone and is in Aristotle and Hippocrates and several times in Galen. But *clavis* was also a curved rod for trundling a hoop (*clavis trochi* of Propertius) or a lever for turning a press (*clavis torculari* of Cato, *De re rustica*). *Clavicula*, perhaps, entered the anatomical vocabulary in one of these senses. It appears in Gerard of Cremona's Latin translation of the *Canon* of Avicenna. Thence it came into the medieval vocabulary.

(105) **Kleides** = lit. keys. See note 104.

(106) *Jugulum* = little yoke, used for clavicle by Celsus. It is a normal term of medieval anatomy. For its relation to the term jugular vein, see note 39.

(107) TARQUWA = clavicle, an Arabic word thus transcribed in Hebrew letters by all three MS translations of the *Canon* of Avicenna. The printed Hebrew Avicenna, on the other hand, uses the biblical SHEKHEM, which appears in the *Fabrica* (1543, p. 167) in place of TARQUWA. The substitution is none too apt, since in the Bible SHEKHEM means *scapula*, but it is in keeping with the hebraizing tendency of the editors of the Hebrew printed Avicenna. The use of TARQUWA here indicates that the Hebrew adviser of Vesalius for the *Tabulae Sex* had access to MS sources. Further, it suggests that he was a different person from the adviser of Vesalius for the Hebrew of the *Fabrica*, who relied almost entirely on one printed source.

(108) *Furcula* as an equivalent of *clavicula* is perhaps an error of Vesalius, since Mundinus leaves no doubt that by *furcula* he means the thoracic basket, that is ribs plus sternum and, perhaps, plus clavicles. Vesalius seems partly to admit his error by his treatment of the word in the *Fabrica*.

(109) In the figure both clavicles are incorrectly placed, and the right is drawn back to front.

C. **Akrōmion** [110], *summus humerus* [111], superior process of scapula, by Galen, in the book *De usu partium*, called **korakoeidēs** from its likeness to a crow's beak [112], ALZEGEM ḤARṬUM *alzegam charton* [113]. The process of this—with the end of which the clavicles articulate by a joint—is properly called **katakleidē**, that is ' at the clavicles ' [114], *rostrum porcinum* [115].

(110) **Akromion** = lit. extremity or top of shoulder. The word is used by Galen in the *De ossibus ad tirones*, Ch. XIV (Kühn, II, 767), one of the most frequently printed of all his works.

He there says that ' certain anatomists declare that in addition to the two bones (scapula and clavicle) which are joined together, there is here a third, found only in human subjects, and this they call the **katakleis** or **akrōmion** '. This discovery is ascribed by Rufus, *De appellatione partium*, Ch. 29, to Eudemus. The ' acromion ', ossified by two centres, unites with the spine of the scapula at about sixteen. In Tab. V and, better, in Tab. VI the suture between acromion and spine of the scapula can be clearly seen. In the *Fabrica* Vesalius rejects the existence of the acromion as a special bone, but emphasizes the acromio-clavicular interarticular cartilage, of which he gives careful figures (*Fabrica*, 1543, pp. 99 and 101). His attention to this cartilage, often absent and seldom well developed, may be his reason for ascribing an error to Galen.

(111) *Summus humerus* is Vesalius' translation of the Greek term. In classical Latin *humerus* has the general significance of ' shoulder '. Vesalius evidently regards the acromion as the highest part of the bony structure of the shoulder. This is hardly the case, since the distal end of the clavicle is higher.

(112) The reference to *De usu part.* is to Book XIII, Ch. 12 (Kühn, IV, 132). **Korakoeidēs** = crowlike, is taken by Vesalius to mean ' like a crow's beak '. But the process is markedly crooked, while a crow's beak is straight. Nor is it good Greek to say ' like a crow ' when one means ' like the beak of a crow '. Vesalius' interpretation seems to be influenced by the Hebrew PI HA-ʿOREBH = *MINQĀR AL-GHURĀB*, which does mean ' crow's beak ' (see note 113). The key to the puzzle lies, perhaps, in the name *AL-GHURĀB* = the crow, given by the Arabs to some part of the ' haunch bone ' (see note 237). The Arab philologists differed as to the exact location of that part (*cf.* Ibn Sīda, al-Mukhaṣṣaṣ, II, 164), a sure sign that the word was foreign. Probably it goes back to some Greek term **korax** or **korakoeidēs**. In view of the general tendency to have the same names for corresponding parts of the upper and lower limb, the name may well have been transferred from some part of the scapula to what appeared to be the corresponding part of the ilium in relation to the joint. The acromion of the Greeks seems to have been that part of the scapula to which Vesalius' ' acromion ' was attached, not the processus itself. The latter may, in fact, have been called by the Greeks **korōnē**, a word which, among other things, means a crow's beak. See note 226.

(113) ALZEGEM ḤARṬUM. This outlandish term for the acromion arises from a series of confusions in Hebrew Avicenna MSS. These we can trace to their origins. ALZEGEM ḤARṬUM contains fragments of two distinct names for the part. The first, ALZEGEM, is a corrupt transcription of *AL-AKHRAM* (see below). The second is ḤARṬUM HA-ʿOREBH = *MINQĀR AL-GHURĀB* = the crow's beak, rendering **korakoeides**. In many MSS the WĔ- = and, which joins the two terms, is misplaced. In the tradition on which Vesalius drew, it had come to stand before HA-ʿOREBH, resulting in ALZEGEM ḤARṬUM WE-HA-ʿOREBH = lit. *alzegem hartum* and the crow. Hence ALZEGEM ḤARṬUM, a grammatically impossible combination, which literally can only mean ' the acromion, a beak ', came to be given as a name for the acromion. In the *Fabrica*, the term appears in an even more corrupt form, taken from the printed Hebrew Avicenna.

AL-AKHRAM = the flat-nosed [bone], seems to represent a popular etymology of some such form as *AKRAM*, transcribing **akrōmion**. It occurs in Arabic as early as the philologist Abu ʿUbaida (729–826), who knows it as *AKHRAM AL-KATIF* = acromion of the scapula (*cf.* Ibn Sīda, al-Mukhaṣṣaṣ, I, 162). This is before the time of the great translators. ALZEGEM and numerous other variants found in Hebrew MSS arose through supplying the wrong diacritic points to الاخرم *'AL-AKHRAM'* in manuscripts where these were absent.

(114) **Akrōmion** is equated with **katakleis** by Galen, *De usu part.* XIII, Ch. 12 (Kühn, IV, 132).

(115) *Rostrum porcinum* = pig's snout, ill fits either acromion or coracoid. It seems highly probable that it is corrupted from *rostrum corvinum* = crow's beak, which is found in the notes on Gerard's translation of the *Almansor* of Rhazes (Leiden ?, 1510).

D. The process of the shoulder-blade on the interior [116] and lower side called **ankyroeidēs** from its likeness to an anchor. Galen often refers to it as **korakoeidēs** and *sigmoeides* [117]. ʿEYN HA-KATHEPH *aijn hacatheph, oculus scapulae* [118].

| *Latin* | **Greek** | HEBREW | *ARABIC* |

(116) The lateral angle of the scapula is much out of drawing and looks as though made from
memory. The coracoid cannot be said to be on the inner and lower side.

(117) The equation **korakoeidēs = ankyroeidēs = sigmoeidēs** is from Galen, *De usu part*. XIII,
Ch. 12 (Kühn, II, 132–3). On **korakoeidēs**, see note 112. ' Anchor-like ' is less inappro-
priate than might be thought. The early Greek anchor was a mere hook which only later
developed prongs on both sides. **Sigmoidēs** is a comparison to the Greek capital sigma,
shaped like our C.

(118) The term 'EYN HA-KATHEPH = *oculus scapulae* = eye of the scapula, is unintelligible
as describing the coracoid process, though it might distantly recall the glenoid cavity. It
is a translation of '*AYN AL-KATIF*. These words are an Arabic innovation found first
as a gloss in Hubaish's version of Galen, *De anat. admin*. (Simon, Vol. I, p. 309, and Vol. II,
p. 224), and apparently intended for the spine of the scapula. Haly Abbas, like Avicenna,
calls this spine *oculus* and gives the doubly nonsensical explanation that ' since there are
no eyes in the back . . . nature has created on each shoulder-blade a spine as a kind of
protection for that part of the thorax '. Arabic uses the term '*AYN* = eye also for the
hind parts of other bones, *e.g.*, '*AYN AL-FAKHIDH* = hind part of the femoral bone
(Ibn Sīda, al-Mukhassas, II, 48). This usage is so strange that we suspect it to have arisen
from an analogy with '*AYN AL-KATIF* (see also ' eye of the knee' = *patella*, note 164).

Now Galen calls the coracoid both ' anchor-like ' and ' sigmoid ' (note 117). These
are intelligible terms for that structure. We suggest, therefore, that the Arabic name
is drawn not from the word '*AYN* = eye, but from the letter '*AYN* = ع, which looks
much like a Greek uncial sigma = c. The proper translation would then be ' the *ayin*
of the shoulder-blade '=' the *sigma* of the shoulder-blade '.

E. *Os pectoris* [119], **sternon** [120], HE-ḤAZEH *hechaseh* [121], *cassos* [122], consists of seven bones, being joined by union rather than by articulation,
like the ribs which are bound to it in the lower part [123]. It is crescent-
shaped (*lunatum*) on either side.

(119) *Os pectoris* is used for the sternum by Mundinus and the medieval anatomists. It is retained
by Berengar (1523).

(120) **Sternon** in its original form = breast, chiefly of males. Thus **osteon sternou** = breast-bone.
Galen cuts out **osteon** and so gives *sternum* its modern significance, *De usu part*. VIII, 2
(Kühn, II, 592).

(121) HE-ḤAZEH = the breast, is biblical. The correct Hebrew term seems to have been
EMṢA' HE-ḤAZEH = the middle of the chest. It thus appears in the printed Hebrew
translation of the *Canon* of Avicenna. In the *Fabrica* it is 'EṢEM HE-ḤAZEH = bone of
chest. Thus to drop the 'EṢEM and to call the sternum HE-ḤAZEH is an exact parallel
with **sternon**. (See notes 120 and 122.)

(122) *Cassos* is from Arabic *AL-QASS*. This, again, has the same history as *sternum*. In classical
Arabic it means breast = *pectus*. Avicenna coined for it the specific anatomical meaning
breast-bone = *sternum*. It appears in Gerard of Cremona's Latin translation of the *Canon*
of Avicenna as *cassum*. Alpago, the Paduan editor of Avicenna's *Canon*, who was contemporary with Vesalius, did not include it in his list of Latin words derived from Arabic.
Perhaps because of this omission Vesalius mistook it for Greek and spelt it *cassos*. There
was no classical word of this form available to Vesalius. Many early Latino-Greek
vocabularies equate *cassus* = **kenos** = empty.

(123) The sternum as depicted by Calcar has seven distinct segments, including xiphisternum.
This error is repeated, though more obscurely, in the *Fabrica* (pp. 86, 87, 164, and perhaps
p. 163). Now there are seven segments in the sternum of the Macaca, but in man all except
the manubrium are united by the 25th year and most of them some years earlier, while even
in children only six segments can be seen. Galen repeatedly ascribes seven segments to
the sternum. Berengar follows Galen's error. The sternum of both the *Tabulae* and the
Fabrica is human and not simian, but the drawings have the ape's anatomy in the background. The error is corrected in one figure of the second edition of the *Fabrica* (p. 107),
but not in the others. Of contemporaries of Vesalius, Massa (1537) describes the sternum

in seven pieces, but Charles Estienne (c. 1538) says 'this bone appears to us never to be formed of more than three pieces, despite that Galen makes out seven'.

(124) *Ab utroqu latere lunatum*, ' lunated on both sides ' a quotation from Celsus (VIII. 1. 15), as is acknowledged in the *Fabrica* (p. 93). Isidore (XVIII, vii, 3) uses *lunatus* to describe a spear-head. The word lingered in anatomy, and James Douglas (1675–1742) still speaks of the sternum as 'lunated' in the 18th century (*Philosophical Transactions*, XXV (1708), 2216).

F. **Xiphoeidēs**, *cartilago ensiformis* (125), by which name also the whole bone is known, ALḤANGRI *alchangri* (126), *ensifoidis* (127), *malum granatum* (128), epiglottal cartilage (129).

(125) **Xiphoeidēs** = *ensiformis* = sword-shaped. The sternum is likened to a sword and specifically to the broad short Roman sword, used for stabbing rather than striking. This is clearly expressed in the traditional term **chondros xiphoeidēs**, used by Galen in the *De usu partium* VI, Ch. 3; Kühn, III, 416. This is rendered *cartilago ensiformis* by Nicholas of Reggio, the fourteenth-century translator of this work from Greek.

(126) ALḤANGRI, *i.e. AL-KHANJARĪ*, from *KHANJAR* = dagger, is used for the xiphoid process by Avicenna and thence passed into Latin in Gerard's translation. Alpago in his index of Arabic words writes: ' *Alchangiari* is derived from *alchangiar*, which is a dagger much in use in Syria. It is broadish, with a sharp point. The cartilage at the end of the thorax, one side of which covers the *os stomachi* (*i.e.*, pylorus), is called *alchangiari* because it is like the point of this sword '.

(127) *Ensifoidis* is a hybrid Latino-Greek word, a mere assimilation of *ensiformis* to **xiphoeidēs**.

(128) In Latin, both classical and medieval, *pomum* or *malum* is a general term for a knob, bulb or blunt end. Throughout the Romance languages the words are used for any fruit resembling an apple. In the Eastern Mediterranean the only apple-like fruit is the pomegranate. This is the *pomum* or *malum granatum*. This term for xiphoid is familiar to the medieval anatomists and is used by Mundinus, who says it ' is made to guard the mouth of the stomach '. (See notes 126 and 343.)

(129) The origin of the extraordinary term ' epiglottal cartilage ' for xiphoid is in Gerard of Cremona's Latin translation of the *Canon* of Avicenna. There we read that ' the lower end of the thorax is continued as a broad cartilaginous bone, of which the extremity is rounded and called *epiglottalis* as resembling the epiglottis '. This is nonsense, for the xiphoid has no resemblance to the epiglottis. Some verbal confusion is evidently involved which had already arisen in the Arabic. We have seen that the xiphoid was called *AL KHANJARĪ* from *KHANJAR*, dagger. From an early date the latter Arabic word was falsely connected with *HANJAR* = throat. The two words would be identical in a normal MS. Graciano, and the printed Hebrew Avicenna translate the word for xiphoid as 'ESEM GĒRONI = throat-bone. (This last term appears also in the *Fabrica*.) Meathi and Lorci point out a similarity of this bone to the Adam's apple. Thus, as often, the discovery of this non-existent similarity came after a linguistic confusion which suggested it !

G. **Brachiōn**; *brachium*, the *humerus* of Celsus and of Caesar (130), ZĒROA' *zeroach* (131), *adjutorium brachii* (132), *aseth* (133). This is smaller than the [corresponding] bone of the lower limb (*tibia*) (134).

(130) In classical usage *humerus* = shoulder and **brachiōn** = upper arm. The Latin term assumed its present meaning in Celsus and the Greek in Pollux. Both these authors were renaissance discoveries. Celsus (first century A.D.) was unknown to the Middle Ages. The *editio princeps* of this work was that of Florence, 1478. A good edition was printed by Giunta at Venice in 1528, but Vesalius had evidently studied the work in the edition of Johannes Caesarius (his Caesar), printed at Hagenau in 1528 with an introduction by Melanchthon. This was the first ' scientific ' edition with notes, variant readings and index. To have selected an out-of-the-way foreign edition from among ten that had appeared (four in Venice) is evidence of critical acumen on the part of either Vesalius or of Guenther. The

Latin **Greek** HEBREW *ARABIC*

text of Pollux (134–192 A.D.), also unknown in the Middle Ages, appeared first at Venice
in 1502, produced by Aldus Manutius. It was at once used by the anatomist Benedictus
in his *Anatomice sive historia corporis humani*, Venice, 1502. In this work of Benedictus
modern anatomical nomenclature may be said to begin.

(131) ZĔROA' is biblical, being mostly applied to the forearm. It is thus employed by Meathi
and Lorci. The printed Hebrew edition of the *Canon* of Avicenna (1491), however, following
Graciano, used it for the upper arm. Doubtless on account of this confusion Vesalius or
Lazarus in the *Fabrica* give the phrase of the printed Hebrew Avicenna QĔNEH HA-
ZĔROA'= shaft of the upper arm.

(132) *Adjutorium brachii* is a redundancy. *Adjutorium* = auxiliary, is a common medieval term
for the humerus. It is a translation of the Arabic *SĀ ID* = cubitus, *lit.* ' supporter '
The wrong application to the humerus is found in Gerard's translation of the *Canon* (I, 1,
V, 18). The confusion must therefore have passed into common usuage before 1180. (See
notes 133 and 139.)

(133) *Aseth* is obscure and is susceptible of two explanations : (*a*) it may stand for the *alseid* of Gerard
of Cremona's Latin translation of the *Canon* of Avicenna =AL-SĀ'ID = forearm (*cf.*
note 132) ; (*b*) it may be for '*ADUD* = upper arm, having passed through Hebrew, where
the word was, as in the transcription in the *Fabrica* (*hasad*), pronounced 'ESED, as if it
were a segolate noun. (' Segolate ' are nouns with the vowel E=*seghol* in the second
syllable, and mostly in the first syllable as well. This is the most frequent pattern of nouns
in Hebrew.)

(134) *Tibia* is a term used loosely by Vesalius. In medieval language it is most often, as here,
equivalent to lower limb as a whole.

H. *Sinus*, dividing the head of the humerus as into two tubercles [135].

I. Groove of the humerus, like a pulley.

(135) *Sinus* is loosely used by Vesalius for any sort of cavity, especially in a solid structure.

K. *Cubitus* [136], **pēchys** [137], HA-QANEH *hakaneh* [138], *asaid* [139], by which
names also this whole part is called [140] ; *ulna* [141], greater focile [142],
ZINĀD 'ELYON *zenad elion* [143]. Its sharp process toward the carpus
(*brachiale*) is called **styloeidēs** [144] (' Styloid process ').

(136) *Cubitus*, normal Latin for forearm.

(137) **Pēchys**, normal Greek for forearm, passing into the meaning of ulna.

(138) HA-QANEH. The word meant originally ' reed '. In the enumeration of the human bones
in the Mishnah (*Oholoth*, I, 8) it is used for the forearm. It is also employed in this sense
by Meathi and Lorci. Graciano, and following him the printed Hebrew translation of
the *Canon* of Avicenna, use instead ZĔROA'. (See note 131.)

(139) *Asaid*=*AL-SĀ'ID* = forearm, appears in medieval works in a great variety of forms : *alsahad*
(Gerard), *alscid*, *absceid*, etc. We have not traced the source of the form used by Vesalius,
which is fairly close to the Arabic.

(140) See notes 131 and 133.

(141) *Ulna* connected with ōlenē = elbow, which English word contains the same root. *Ulna* is not
used in our sense by classical or medieval writers, with whom it normally means elbow.
The modern sense of ulna seems to have come in about the time of Vesalius, for Charles
Estienne uses it and says it was employed by ' the Latins '. It is perhaps a new Parisian
usage, possibly of Sylvius.

(142) *Focile* is a very interesting transferred use of late Latin *focile* = tinder for lighting fire (*cp.*
classical *focus*, *foculus* = hearth). The Arabic for the bones of the forearm is *ZAND* =
one of a pair of sticks for producing fire by friction. This Gerard rendered by *focile* as
the nearest Latin equivalent in his translation of Avicenna's *Canon*. The word *focile* thus
passed into medieval Latin, into the Romance languages, and into English, the greater
focile or ulna being distinguished from the lesser focile or radius.

(143) *ZINĀD 'ELYON*= lit. ' upper *zinad* '. The word is a corruption of *ZAND*, perhaps the
plural used for the singular. (See note 142.) There is confusion here, for the ulna should

be the lower of the two forearm bones. In any event the two bones of the forearm are very badly drawn, and those of the right arm have the bones of a left hand attached to them.

(144) **Styloeides** = rod-shaped is from Galen.

L. *Radius* [145], **kerkis** [146], ZINĀD TAHTON *zenad thachton* [147] ; lesser focile [148] of the brachium [149].

(145) *Radius* is from Celsus. The first modern medical writer to use it was Benedictus (1502). It means primarily the spoke of a wheel, our word ray. Celsus explains *Radius, quem Graeci* **kerkida** *vocant, superior breviorque, et primo tenuior . . . Cubitus inferior longiorque, et primo plenior* [radio], VIII, Ch. 1, ' The radius, which the Greeks call **kerkis**, is upper, shorter and at first more slender . . . The cubitus [=ulna] is lower, longer and at first stouter [than the radius] '. Benedictus copies Celsus. Vesalius doubtless took the word from Guenther's *Inst. anat.*, Basel, 1536. The radius is much out of drawing in both limbs in this plate, though better in the subsequent ones.

(146) **Kerkis** = primarily a weaver's shuttle and afterwards a slender rod of any kind. Galen uses it to describe the radius, *De usu part.* II, Ch. 2 ; Kühn, III, 92, and elsewhere.

(147) ZINĀD TAHTON, lit. lower ZINĀD or ZAND. (See notes 142 and 143.) It should be the upper of the two bones.

(148) On *focile*, see note 142.

(149) *Brachium* is used confusedly by Vesalius. (See note 130.)

M. [Though marked in figure is not described in the text, but see under L. It indicates the styloid process and separate epiphysis of the ulna, which unite with the shaft about the 20th year.]

N. *Brachiale* [150], **karpos** [151], RESEGH *reseg, raseta, rascha* [152], consists of eight dissimilar bones, set out in two rows, three in the upper, four in the lower. These, as the plate shows, form a figure which has both an inner hollow and an outer bulge. Their number is uncertain according to Celsus [153].

(150) *Brachiale* is the term for our carpus in Gerard of Cremona's Latin translation of the *Canon* of Avicenna. It is adopted by Guenther in his *Inst. anat.*, 1536, whence Vesalius doubtless takes it.

(151) In Aristotle **karpos** means simply wrist. In Galen, *De usu part.* III, Ch. 8 ; Kühn III, 203, and often, it comes to mean the bones of the wrist. Thus it evolves in the same way as the word *humerus*—from the part in general to the bones of the part. (See note 127.)

(152) *Raseta* and its innumerable variants, a few of which are given here by Vesalius, was the usual medieval term for carpus. It was introduced as *rascha* by Stephen of Pisa in his version of Haly Abbas and as *raseta* by Gerard in his Avicenna. It is from *RUSGH*, used by the Arabic physicians to designate the carpus, but meaning primarily the ' pastern-joint of an ungulate '. From it is derived a verb, *RASAGHA*, ' to tether an animal by the pastern-joint '. The Hebrew pronunciation RESEGH is due to assimilation of the foreign word to the Hebrew segolate pattern. (See note 133.) It is strange that the Hebrew translators should have introduced this Arabic word, since Hebrew possesses several words for the carpus —such as ASSIL, PEREQ, etc. It is thus significant that Gerard, by his choice of the form *raseta*, was evidently influenced by a Hebrew speaking interpreter. It has long been suspected that his work, conducted at Toledo, the main Jewish centre in Spain, was carried on with Jewish help.

(153) Celsus, Bk. VIII, Ch. 1, 22, ' the first part of the palm is made up of many minute bones, of which the number *is not uncertain* '. Thus all early editions. Vesalius has corrected a redundant negative.

Latin	**Greek**	HEBREW	*ARABIC*

E

O. **Metakarpion** (154), *palma, pecten* (155), MASREQ *masrek* (156), *postbrachiale*, is formed of four bones, as Galen says, and not of five as most others reckon (157).

(154) **Metakarpion** is frequently used by Galen in *De usu part.* and adopted by Berengar in his *Isagogae*, Bologna, 1523, and by Guenther in his *Inst. anat.*, Basel, 1536. The term was not used in the Middle Ages.

(155) *Pecten* = comb is a common and obvious medieval expression, dating at least from Gerard of Cremona, for the skeleton of the hand. It translates *MISHT*, which in turn is the classical Greek **kteis** and **ktenion**. *Pecten* is one of the few medieval terms retained by the classical purist Guenther both in his translation of the pseudo-Galenic *Introductio seu Medicus*, Paris, 1528, and in the *Inst. anat.*, 1536.

(156) MASREQ is Mishnaic Hebrew for comb. It translates Arabic *MISHT*.

(157) Galen, *De usu part.* III, Ch. 8 (Kühn, III, 203). The ' others ' are Eudemus, as Galen says, and Celsus.

P. **Daktyloi**, digits, ESBA'OTH *esbaoth* (158), formed each of three bones, 'the more proximal ' internode ' fitting into the ' sinus ' of the next (159).

(158) ESBA'OTH, singular ESBA' = biblical for fingers. In the text the Hebrew letters are printed the wrong way round, from left to right !

(159) *Sinus.* (See note 135.)

Q. **Mylē** (160), **epigonatis** (161), *patella* (162), *rotula genu* (163), MAGHEN HA-ARKUBBAH *magen harcubach* (164), shield of the knee, *aresfatu* (165), a round bone like a short shield (166).

(160) **Mylē** = mill, hence millstone, hence knee-cap. The last usage is Hippocratic and Aristotelian. Galen, *De usu part.*, III, Ch. 15 (Kühn, III, 253) and elsewhere.

(161) **Epigonatis** = upon the knee. A rare word used for knee-cap in the Hippocratic *De articulis*, and by Galen in his *Comment. in Hipp. de articulis*, III, Ch. 60 (Kühn, XVIII B, 626. In the Aldine edition of the *Opera*, in Greek, Venice, 1525, Vol. V). It is here merely empty learning.

(162) *Patella* is not a medieval anatomic term. It is classical Latin for a small dish of the type used for votive offerings. It is used anatomically by Celsus and adopted by Benedictus, *Anatomica*, 1502, and Guenther, *Inst. anat.*, 1536.

(163) *Rotula genu*=knee millstone is the usual medieval anatomical term for *patella*. *Rotula* doubtless translates **mylē**. In Italian the patella is still called 'rotella del ginocchio'. Gerard (c. 1170) gives *mola* as a popular term for the patella (*oculus genu*) in his translation of the *Almansor* of Rhazes.

(164) MAGHEN HA-ARKUBBAH = shield of the knee. ARKUBBAH is Mishnaic Hebrew, taken from Aramaic, the biblical form being BEREKH. This word is used by the Avicenna translators Meathi and Lorci, while Graciano and the printed Hebrew Avicenna use 'EYN HA-ARKUBBAH = eye of the knee, a closer rendering of '*AYN AL-RUKBA*. The *Fabrica* gives the two terms side by side. (See note 118.)

(165) *Aresfatu—AL-RADAFĀT*, plural of *AL-RADAFA* or *AL-RADFA* = patella. The word came into Latin in many different forms, derived partly from the singular (*alresafe* in Alpago's glossary, etc.), partly from the plural (*alrasefati* in the glossary appended to the older Avicenna editions). It is confused with *rasga* = metacarpus in the Latin Avicenna edited by Alpago.

(166) We have not traced this phrase. In Hebrew the bone is called a shield. (See note 164.)

R. **Astragalos** (167), *talus* (168), QARSUL *karsul* (169), *os balistae* (170), *cavilla* (171), *chahab* (172), *alsochi* (173). Some nowadays have wrongly changed it to *malleolus* (174).

(167) **Astragalos** = ankle-bone. Those of sheep and goats were used as dice, and thus *astragalus* became the name for a dice. The dice of the ancients had only four flat sides, the other two being left rounded.

(168) *Talus* is normal Latin for the astragalus. Confusion arises from its also being used for the heel, that is for the calcaneum. *cp.* French talon.

(169) QARSUL. The biblical word, perhaps, means hock. Its plural is translated by the A.V. as "feet" in II Samuel, xxii, 37 and Psalm xviii, 36, but the Greek translators had rendered it by **sphyron** = heel. Of the Hebrew Avicenna translators only Graciano avoids this word, using instead a transliteration of *KA'B*.

(170) *Os balistae* seems to have been taken by Vesalius from his forerunner at Padua, Gabriele de Zerbi (died 1505), editions of whose *Anatomia corporis humani et singulorum illius membrorum* appeared at Venice in 1502 and 1533 (edition of 1533, folio 181). In this work the *os balistae* is not so named as a projectile from the ancient ballista, the siege-engine, but rather as part of the arcubalista or arquebus, the successor of the crossbow and predecessor of the musket. Charles Estienne calls attention to the connection of the term with the arquebus and indicates that it is a modern introduction.

(171) *Cavilla* is Latin for a joke. As an anatomical term, however, it has another derivation. It is first found in the Latin version of Haly Abbas (Lyons edition, 1523, *Theorice*, Lib. II, Ch. 8). It is from *cavicula*=wedge heel, *cp.* Italian 'caviglia', itself a modification of classical *clavicula*. Hyrtl unnecessarily derives it from a non-existing *QUBILA* = wedge. (*Cp.* note 220.)

(172) *Chahab* =*KA'B*. It occurs in this form in the work of Berengar (1523). In Mundinus (1316) it is *caib*, and in Albertus Magnus (*c.* 1250) *schib*. *KA'B* is used for the astragalus by the medical writers, but originally seems to have meant the malleoli (see the description in Ibn Sīda, al-Mukhaṣṣaṣ, II, 56).

(173) *Alsochi.* The word does not occur in the Latin writers we have consulted. However, in the Hebrew translation of Rhazes' *Almansor* the bone is called אלשוב ALSHOBH, which looks like corruption from אלשוכי = Arabic *AL-SHAWKĪ* = the spur-bone. The Latin (ed. of Leiden, 1510) has *aschib*, which is even more hopelessly corrupted.

(174) *Malleolus* = little hammer. The passage on the wrong naming of the astragalus as malleolus is copied from Galen, *Com. in Hipp. de med. off.* I, Ch. 6 (Kühn, XVIII B, 756). But the error was made also in the sixteenth century. In 1536 Nicholas Massa of Padua and Venice published his *Liber introductorius anatomiae sive dissectionis corporis humani*, in which he speaks of the astragalus as the *malleus* = hammer. (See also note 264.)

S. Boat-shaped, **skaphoeides**, *naviculare* [175], ṢORQI *zorki* [176].

(175) **Skaphoeidēs** = *naviculare* = boat-shaped, used by Galen, *De usu part.* III, 6 (Kühn, III, 195). An appropriate name.

(176) ṢORQI is *ZAURAQĪ*, from *ZAURAQ*, a small boat. The rendering of Arabic *Z* by the Hebrew letter ṢADE is unusual and seems taken from current medical speech. The three Hebrew Avicenna translators write ALZORQI. The printed Hebrew Avicenna has instead the two words SĒPHINAH QĔṬANNAH = small ship, and 'ESEM HA-SHAWEH = the equal bone. Both Hebrew terms are reproduced in the *Fabrica*. The latter name seems to be due to a misinterpretation of *AL-'AZM AL-MUQAWWAS* = the curved bone, which was confused with *MUQAYYAS* = equal.

T. Tarsos [177], RESEGH *reseg, raseta pedis* [178], consists of four bones of which the largest is placed outermost and called **kyboeides** because of its dice-like shape, *tesserae os* [179], ṬARDI *thardii* [180], *exagonon* [181], *grandinosum* [182], *nerdi* [183]. The remaining three lack names but are called by some **kalkoeidē** [184]. We have twice seen a right foot to have an additional [bone].

(177) **Tarsos** is used by Galen, *De usu part.* Lib. III, Ch. 8 (Kühn, III, 200), and elsewhere in the modern anatomical sense. Its primary anatomical meaning seems to be the flat of the foot.

(178) *Reseg, raseta*, etc. (See note 149.) In Arabic, as with us, the nomenclature of hand and foot are closely parallel.

Latin	**Greek**	HEBREW	*ARABIC*

(179) **Kyboeides**, Galen, *De usu part.* III, 6 : Kühn, III, 199 *et passim*, implies a bone bounded by rectangular sides.

(180) *Thardii* in the *Fabrica* becomes *chardii*, but see below. The Hebrew letters stand for $\bar{N}ARD\bar{I}$, from *NARD* = backgammon-stone. The Arabic نردى was misread as تردى *TARDĪ*, which appears as תרדיי in the translators Meathi and Lorci. The form in the text doubtless arose by oral transmission, ת (T) and ט (Ṭ) not being distinguished in Italian Hebrew pronunciation. Another corruption of the Arabic is بردى *BARDI*, which appears in the printed Hebrew Avicenna as ברדיי. In that book the ב is not quite clear in the print. The Hebrew assistant of Vesalius for the *Fabrica*, Lazarus de Frigeis, read the word as KARDI. Of the translators, only Graciano understood the Arabic word. He translated it as 'EṢEM HA-DOMEH LE-'AṢMOTH HA-QUBHYA = bone similar to the bones of the game ' *qubhya* '. This last word = **kybeia** = ' dice '. The term *nerdi* occurs in Gerard's Avicenna glossed *hexagonon*. On it Alpago makes this amusing comment : ' It is a bone placed at the extremity of the pecten (note 155) of the foot toward the outer part. . . . It is called *nerdi* because its form is like a six-sided piece (*taxillus*) with which the ancients used to play. These pieces were called *nerdi* from their inventor, the distinguished philosopher Nerdi '. *NARD* is in fact backgammon, and the story is a confusion of the famous legend of the invention of chess.

(181) **Exagonon** for **hexagonon** = six-sided. From Gerard's t·anslation of the *Canon* (I, 1 ; V, 1, 30).

(182) *Grandinosum* = hail-like. This does not refer to any similarity of the bone with the six-sided ice-crystal, but is a translation of BARDI (for NARDI) which appears in the printed Hebrew Avicenna (*cf.* note 180). Both in Hebrew and Arabic BARAD means hail. Since Vesalius' Jewish adviser gave him the form ṬARDI, the *grandinosum* cannot be due to him, but must be so old that its Semitic origin was not apparent. Benedictus (1502) calls it *grand.neum*.

(183) *Nerdi.* (See note 180).

(184) **Kalkoeidē** should be **chalkoeidē**. This word is applied to the cuboid in the pseudo-Galenic *Introductio seu medicus*, Ch. XIII (Kühn, XIV, 725), but it is probably older than Galen, since it occurs in a papyrus of the 2nd to 3rd century A.D. as **[cha]lkoeides ostoun,** after the **astragalos** (H. J. M. Milne, *Literary Papyri in the British Museum*, London, 1927, no. 167, line 16). Its etymology is enigmatic. The obvious derivation, from **chalkos** = copper, does not make sense, as the bone is not copper-coloured, but there is no other suitable Greek word, unless it be **chalchē, chalkē, kalchē** = rosette of a column. The occurrence in the 2nd to 3rd century precludes any connection with *calculus* = little stone used for games. In any event, Vesalius would derive the word not from the Galenic book but from Guenther's version of it (Paris, 1528, p. 21).

V. The *planta, planum,* **pedion** [185], *pecten* [186], MASREQ *masrek* [187] is formed of five bones. After it come the digits of the foot, [marked] X, which all consist of three ' internodal ' bones except the great toe which, alone among the others, is formed of two bones.

The ossicle at the first joint of the great toe is one of the sesamoid [188] bones, and in that particular position we have observed two in each foot [189].

The bones of the human body some reckon as 248, others as some other number. Omitting the hyoid, which is compounded almost into one by synchondral union of six little bones, and omitting also the sesamoids, I reckon there are 246, as comprised in the distich of the plate which follows [190].

(185) **Pedion** is used for metatarsus by both Galen and Pollux.

(186) *Pecten.* (See note 155.)

(187) MASREQ. (See note 156.)

(188) *Sesamina* is adapted from **sēsamoeidē osta**, used by Galen, *De ossibus* (Kühn, II, 778). Perhaps the Latin form is due to the Arabic $SIMSIM\bar{A}NIYYA$.

(189) The sesamoid of the big toe is known as the *Os Vesalianum*. This is the first reference to it.

(190) Medieval writers attach great importance to the **exact** number of the various structures in
     the human body.   Many distichs and longer collections of verse were constructed to assist
     the learner to remember such useless ' Pythagorean ' pieces of information.   These methods,
     quite senseless to our way of thinking, were part of the mental furniture of the men of the
     day.   It must be remembered that the key-thought of the age from Galen onward was
     the Divine Purpose.   Each structure in man's body was specially designed by God for a
     special purpose.   These numbers were thus summaries of the supposed knowledge of these
     purposes.   As regards the number itself, the opinion of medieval and renaissance physicians
     varied widely.   That mentioned by Vesalius, 248, coincides with that given by Avicenna
     (I, 1 ; V, 30), by Albucasis in his *Theorice* (IV, 1), and by the Mishnah *Oholoth* (I, 8).   The
     belief is thus at least as old as the 2nd century A.D., and was evidently a floating tradition.
     The Greek physicians give no number for the bones.

TABULA V.

Portrayal of lateral aspect of **skeleton** [191].   Bones of the **kranion**, of the
calvaria [192], of the bone of the head, QĔDHERATH HA-MOAH *kadroth
hamuach* [193], *olla capitis* [194], *asoan* [195].

(191) *Skeleton* spelt in Greek letters, like many current anatomic terms in the *Tabulae*.   These
     terms, very familiar to us, were just coming into use in the sixteenth century.   They were
     therefore still presented as Greek words, much as we now sometimes print in italics many
     common French words, such as *rôle, à propos*, and so on.   But it must not be inferred from
     this use of Greek letters and words that medical students, or even medical professors, of
     the period had any facility in Greek.   With very rare exceptions they knew little but the
     letters.   The knowledge of Greek words was part of the mystery of the art of medicine,
     much as are the *termini technici* of modern medicine.   To interpret the use of such catch-
     phrases in terms of scholarship is to misunderstand the situation.

(192) Vesalius follows the correct usage.   The modern *calvarium* is justified neither by the classics
     nor by the Vulgate.

(193) QĔDHERATH HA-MOAH = bowl of the brain.   QĔDHERAH=earthenware bowl is Mishnaic
     Hebrew.   MOAH occurs in the Bible but once, with the meaning ' marrow '.   In the Mishnah
     MOAH = brain.   The form indicated by Vesalius' transcription, QADHRUTH HA-MOAH,
     is due to some mistaken etymology, as if the phrase meant ' cover of the brain '.   Hebrew
     possesses a word for the cranium in the biblical GULGOLETH (*cp.* Golgotha).   This word
     is used by Graciano and the printed Hebrew Avicenna and appears in the *Fabrica*.   The
     Arabic word is *QIHF*, which evokes no association with ' bowl ', nor is there an Arabic
     word suggesting the Hebrew phrase here used.   Presumably QĔDHERATH HA-MOAH
     translates the *testa* = potsherd, pot, and also = head, cranium, which is also found in late
     Latin (*Corpus Glossariorum Latinorum*, IV, 291 ; V, 526, 581).   It would thus be part of
     the professional jargon of the Italian Jewish physicians.   It must have arisen early, as it
     was used by Avicenna's translator, Meathi, *c.* 1279.   In discussing these terms for the
     calvarium we cannot but recall the anatomical interpretation of the exquisite passage in
     the Book of Ecclesiastes (xii, 5–7) : ' When the grasshopper shall be a burden . . . or ever
     the silver cord be loosed, or the golden bowl (GULLAH) be broken . . . and the dust return
     to the earth as it was and the spirit to God who gave it '.

(194) *Olla capitis* = cist of the head : *olla* = a pot or closed vessel, especially for containing the
     ashes of the dead.   The application to the skull is fanciful.   It is a translation of the Hebrew,
     as is shown from the *Fabrica* (p. 166), where the fuller Hebrew term is translated *theca et
     olla capitis*.

(195) *Asoan* is for *AL-SHU'ŪN* = the sutures, singular *SHA'N*.   We have not traced whence
     Vesalius drew the word.   *Soonia* is found for cranium in Constantine Africanus (*c.* 1070).

| *Latin* | **Greek** | HEBREW | *ARABIC* |

A. Two bones of the **bregma,** of the **koryphē** [196], of the *sinciput* [197], of the vertex, *parietalia* [198]. The passage [concerning these] in Avicenna in both the Latin and Arabic version is wrong [199].

(196) Aristotle, *Historia animalium* (491 *a*, 31 ff.) : ' The front portion of the **kranion** is called **bregma**. It is the last of all the [head] bones to acquire solidity. [This refers to the anterior fontanelle.] The hinder portion is called **inion**. That between **bregma** and **inion** is the **koryphē** ' [= crown : Vesalius prints **korēphē**]. Aristotle fails to distinguish between the bones and the regions of the head, and a similar confusion extends through much of ancient, medieval and renaissance anatomical nomenclature.

(197) The curious form *sinciput* is a very early classical contraction for *semi-caput* (Plautus), used for the head and brain in general. It first appears in medieval Latin, in the sense in which Vesalius uses it, in Gerard of Cremona's Latin translation of the *Canon* of Avicenna. It is not used by Mundinus, but is adopted by Benedictus (1502). For *occiput*, see note 209.

(198) *Parietalia* = wall bones, has passed into modern anatomical nomenclature, though it was dropped by Vesalius in the *Fabrica*. We have not traced it further back than Dryander's *Anatomiae . . . pars prior*, Marburg, 1537. The medieval name for these bones, used by Mundinus, is *parietes* = walls.

(199) The reference suggests that Vesalius himself had access to the Arabic. This is certainly not the case. In fact, Avicenna (I, 1 ; V, 1 ; 2) does not mention the parietal bones at all, as he only speaks about the sutures, not about the bones of the cranium. We learn the Arabic name of these bones from Haly Abbas, *'AZMĀ AL-YĀFŪKH* = ' the two bones of the vertex '.

B. Two bones, one at each ear, [bones] of the **krotaphoi** [200], of the temples, of the ears. Of each there are [these] parts ; processus **belonoeidēs** [201], that is like a bodkin or dart (' styloid ') [202], *processus mamillaris* (' mastoid ') and the process of the jugal bone [203] (' zygomatic ').

(200) **Krotaphoi** is a general rather than a technical term for the temples. It is seldom used by Galen, and its introduction here is of no practical use.

(201) **Belonoeidēs** = dart-like ; **belonē** = a needle. The word is used by Galen, *De usu part.* VII, 19 (Kühn, III, 592), and elsewhere.

(202) Vesalius gives the Greek, but not the Hebrew, names for the styloid and mastoid. In the *Fabrica* both receive Hebrew names, though neither is named in the printed Hebrew *Canon* of Avicenna. This accords with the view that he had a Hebrew assistant for the *Tabulae sex* who was different from Lazarus de Frigeis, who helped him with the *Fabrica*.

(203) *Mamillaris* is the translation by Vesalius of Galen's **mastoeidēs**, just as *jugale* is of **zygōmatos**. Neither *mamillaris* nor *jugale* gained currency.

C. Bone of the **metopon** [204], of the forehead, 'EṢEM HA-MEṢAH *etsem hametzah* [205], *coronale* [206]. Sometimes, on account of a straight suture extending to the nose, it appears double. Some falsely consider that this is always the case in women [207].

(204) **Metopon** is a literary, not a scientific, term, with usage something like our word ' brow '. Galen, *De ossibus* (Kühn, II, 745), calls the frontal bone **to kata metōpon** = the [bone] *at the brow*.

(205) 'EṢEM HA-MEṢAH = *os frontis*, from the biblical MEṢAH, ' forehead '.

(206) *Coronale* is a confusion of bone and suture.

(207) Aristotle, *Historia animalium* (491 *b*, 2 ff.) : ' The skull has sutures ; in women, one circular ; in men, three meeting together. Some male skulls are devoid of sutures.' There is a similar passage in *Historia animalium* (516 *a*, 15 ff.) and *De partibus animalium* (653 *b*, i ff.), which continues with the well-known passage on the crocodile's jaw (see note 226). It is difficult to understand how the idea arose that women had sutures in the skull different from those of men. The sutures were of special interest as early as Herodotus (IX, 83).

D. The one bone of the **inion** [208], of the occiput [209], of the 'OREPH *oreph* [210], [*os*] *cunei* [211]. In it is the *foramen maximum*, through which the spinal medulla emerges.

(208) ' The one bone of the **inion** ', that is to say, one as compared to the paired bones. On **inion** see note 196.

(209) *Occiput* is late in classical Latin (Persius and Ausonius). More common is *occipitium* (=*ob-caput*), which is used by Celsus and adopted by Benedictus. By the fourteenth century the term had entered the Italian, French and English vernaculars. On the kindred *sinciput*, see note 197. The common medieval terms for occiput and sinciput were *puppis* and *prora* (poop and prow), introduced into anatomy by Constantine (*c.* 1070) and Stephen of Pisa (1127).

(210) 'OREPH (in text wrongly OR) = back of head (*cf.* p. lxxx) Avicenna and all his translators call the occiput '*AZM AL-RA'S* = 'ESEM HA-ROSH = the head bone. These terms appear to be merely contractions of an older phrase, used by Haly Abbas, '*AZM MU'AKHKHAR AL-RA'S* = bone of the back of the head = **ostoun iniou**. 'ESEM HA-'OREPH, as it appears in its full form in the *Fabrica*, must therefore be derived from Haly Abbas or one of the other older Arabic authors, and not from Avicenna.

(211) The term *os cunei* = wedge bone, is applied by Vesalius both to the occipital and to the sphenoid bone. We cannot trace its use for the occipital. It is hardly likely to be connected with *QUNNA* (also *QUMMA*, *QULLA*) = occiput, a word not found in medical writings (Ibn Sīda, al-Mukhassas, I, 55).

E. *Ossa* **zygōmata**, *jugalia* or **zygoeidē** [212], ZUGH *zog* [213], [*ossa*] *paris* [214]. On either side one, consisting of processes of two bones [meeting] and on that account they have no proper outlines. (Zygomatic arch.)

(212) **Zygon** = *jugum* = yoke. Galen uses both **zygoeidēs** (Kühn, XIV, 721) and **zygōma** (Kühn, II, 437, 746). The term *jugale* was, in the time of Vesalius, a neologism of the Paduan school, taken from Celsus (Lib. IV, Ch. 1). In Balamio's translation of Galen's *De ossibus* (1541) we read *Zygoma a nostris jugale appellatur*, ' the zygoma, called jugal by our colleagues '. Its first traceable use is by G. Valla of Venice in his *De humani corporis partibus opusculum* (1527), who writes : *jugale appellari potest, quia Graeci zygodei alii zygomata appellant.* Since Valla was printed in the same volume as the first Basel edition of Guenther's *Inst. Anat.* (1536), Vesalius certainly knew this passage. The term ' jugal ' was adopted by Charles Estienne (*c.* 1538).

(213) ZUGH = in Mishnah yoke of oxen = pair. It is a loan word from Greek, as is *ZAUJ*. ZUGH means only one **zygōma**, so that strictly the translation of Vesalius, *ossa jugalia*, is wrong. The transcription *zog* is given also in the *Fabrica*. *Zog* would be the normal Hebrew pronunciation of *ZAUJ*, which again suggests an Arabic flavouring to Italian Hebrew medical terminology.

(214) *Ossa paris* is from Gerard of Cremona's Latin *Canon* (Book I, Fen 1, Doct. V, 1, Ch. 3), and a literal translation of the Arabic *AL-'AZM AL-ZAUJĪ* = lit. yoke bone, pair bone. The significance is not that the bone is paired, for that would apply to most bones, but that it forms the arch by meeting its fellow forming the zygomatic arch (= zygomatic bone plus zygomatic process of temporal bone).

F. *Os* **sphenoeides**, *cuneiforme* [215], *basilare* [216], sometimes from its manifold form [called] **polymorphon** [217], MOSHABH HA-MOAH *moschab hamoach* [218], [*os*] *colatorii* [219], *cavilla* [220]. This [bone], between the superior maxillae, is usually said to be numbered fifteenth. For there are six which lie around the eye-socket, two very large ones containing the malars and the alveoli of the molar teeth [221] ; two of the nose ; two holding the incisory teeth ; two at the posterior end of the palate containing the openings of the nares [222] ; and the *os cuneiforme*

Latin   **Greek**   HEBREW   *ARABIC*

which has just been mentioned before all these [223]. (Unless perchance
you prefer to call them eight bones or, following the opinion of certain
Greeks, twelve, according as you either include or omit certain small
sutures, commissures and junctions [224]).

(215) **Sphenoeides** = wedge-shaped = *cuneiforme* is the name given by Galen, *De anat. admin.*
(Kühn, II, 752). Vesalius is the first to render this as *cuneiforme*. (But see note
211.)

(216) *Os basilare* in medieval nomenclature should normally be rendered ' base of the skull '. This
is its meaning in Gerard of Cremona's Latin Avicenna and in Mundinus. *Os basilare* here
is a misunderstanding by Vesalius. It is probably assimilated to the term *os baxillare*.
This is from a rare Latin word, *paxillus*, a peg or wedge, and thus equivalent to *cuneus*.
Vesalius in the *Fabrica* (p. 167) introduces *baxillare*. (See note 218.)

(217) The term **polymorphon** is used in the pseudo-Galenic *Introductio seu medicus*, Ch. 12 (Kühn,
XIV, 720, 721). It was used by Guenther in his translation of the work (editio princeps,
Paris, 1520, fol. 19 v). It seems unknown elsewhere.

(218) MOSHABH HA-MOAH = lit. ' resting-place of the brain ', from YASHOBH = to sit.
This word, employed by Lorci, Meathi, and the printed Hebrew Avicenna, is a poor translation
of the Arabic *QĀ'IDAT AL-DIMĀGH*=basis or pedestal of the brain. The Hebrew
translators, not knowing the meaning of *QĀ'IDA*, guessed at something connected with
' to sit ', the meaning of the verb *QA'ADA*, from which it is derived. Indeed, both
Lorci and Meathi add an alternative translation, KAN HA-MOAH, which is better, but
was dropped in the printed Hebrew edition. However, *QĀ'IDAT AL-DIMĀGH* is not
the real name of this bone at all, which should be *AL-WATADI* = the tent-peg-shaped = the
sphenoid ='the *alguateda* of the medieval Latins. This corresponds to the Hebrew
HA-YĔTHEDHI, appearing as HA-YITHRI in the printed Avicenna. This word was
overlooked by Lazarus in the *Fabrica*. The real meaning of MOSHABH HA-MOAH can
be seen from the description of Haly Abbas (de Koning, p. 112): ' . . . l'os sphénoide
(*AL-SHABĪH BIL-WATAD*) c'est un os réuni à l'os postérieur de la tête (os occipitale,
'AZM MU'AKHKHAR AL-RA'S, see note 210) à l'endroit appelé base du crane (QĀ 'IDAT
AL-RA'S) '. MOSHABH HA-MOAH is thus the basilar part of the occipital. Confusion
between sphenoid bone and occipital bone is thus of early origin. Vesalius merely carries
on the tradition.

(219) ' Bone of the *colatorium* '. The word *colatorium*, a normal medieval anatomical term,
involves a special physiological conception. It is barbarously formed from Latin *colare* =
to strain, filter, purify. Mundinus wrote : ' In the fore-part of the brain are two *carunculi*
[= olfactory lobes] covered merely by the fine *pia mater*, for in man they are not liable to
exposure. The brain, being cold and damp, is strengthened and fortified by odours which
can dry and warm a relaxed brain. Wherefore these parts are in the *colatorium nasi*, whence
vapours, passing through the porous parts of the nose and transformed into odours, are
sent up into the brain '. The *colatorium nasi* is the cribriform plate of the ethmoid.

(220) *Cavilla* is taken from Stephen of Pisa's translation of Haly Abbas, where it has the same name
as the astragalus (see note 171). It may go back to the Romance word for a wedge, as
does the latter, or it may be a wrong application of *QABĪLA* = cranial bone in general,
a word not found in Arabic medical literature, but used in poetry.

(221) This confused passage suggests that the malar and the superior maxilla on each side should
be treated as one very large bone. In this respect Vesalius is simply following Avicenna
(I, 1 ; V, 1 ; 4 : see De Koning, p. 462.)

(222) This system of enumeration is taken from Avicenna. (See De Koning, p. 119.)

(223) The *sphenoid* is described in modern anatomical text-books as articulating with all fifteen
bones of the skull.

(224) With these omissions the sphenoid articulates with 4 single and 4 paired bones. The
' certain Greeks ' is merely a phrase copied from Galen, *De ossibus*, Ch. IV (Kühn, II, 751).
It would therefore be a mistake to consider that it implies deep knowledge of Greek literature
on the part of Vesalius.

G. The two bones of the lower jaw, the anterior part being very firmly joined together by fusion. I am not sure that in the human subject we are wrong to call them one, following Celsus, for I have found it almost impossible to separate them even after prolonged boiling, and to a saw no part is more resistant of separation than their junction [225].

(225) The phrase 'in the human subject' (*in hominibus*) shows that Vesalius was conscious of anatomical differences between men and animals. In such apes as he was likely to examine, the mandibular synosteosis is a good deal less firm than in man, and in fairly young forms can be loosened by boiling. Galen, *De ossibus*, Ch. 6 (Kühn, II, 754), describes the lower jaw as divisible into two without mentioning that it is that of an animal.

H. Korōnon [226].

(226) **Korōnon** should be **korōnē**. This word is used by Galen, *De usu part.* XI, Ch. 20 (Kühn, III, 937), for the coronoid process of the mandible (see note 112). It appears as *corona* in the fourteenth-century translation of the work by Nicholas of Reggio. The term refers neither to the likeness to a crown nor to a crow's beak, but to the **korōnē** or end of the shaft of a plough (*cp.* Pollux, *Onomasticon*, I, 252).

I. Tubercle and neck of the inferior maxilla. These only are said to be movable in all animals except the crocodile [227].

(227) The first occurrence of this odd statement concerning the crocodile is in Herodotus (II, 68): 'He does not move his lower jaw, but brings the upper toward the lower, being in this unlike all other creatures'. The general description of the crocodile by Herodotus is good, and even this statement is quite colourably true. In Africa basking crocodiles rest the lower jaw on the ground, raising the upper jaw together with the skull and occasionally snapping it down again. The statement is repeated by Aristotle in the *Hist. anim.* (492 *b*, 23 ; 516 *a*, 24) and the *De part. anim.* (660 *b*, 27, and 691 *b*, 5). In the latter work Aristotle comes near explaining what seems an anatomical absurdity.

K. Two processes of the ulna, of which the posterior is called **ōlekranon** [228]. They have in the middle a notch (*sinus*) having a likeness to the ancient Greek letter sigma, our C [229].

(228) The letter K is placed clearly on the olecranon process in both arms. The passage is an abbreviation from Nicholas of Reggio's translation of Galen's *De usu part.* II, Ch. 14 (Kühn, III, 142). **Ōlekranon** (=**ōlenēs kranion** = point of the elbow) occurs in the modern anatomical sense in this passage of Galen : **ōle-nē** = *el-bow* = *ul-na*, are etymological equivalents. (See note 141.)

(229) The comparison of the notch at the proximal end of the ulna to the Greek capital letter sigma, the 'sigmoid notch' of modern anatomists, is taken from Galen's *De usu part.* XIII, Ch. 12 (Kühn, IV, 133). There he compares several structures to this C-like letter, for example the coracoid.

Ribs, **pleurai**, ṢĒLA'OTH *tzelaoth* [230], twenty-four both in men and women, namely twelve on each side. Of these, seven are joined at either end with the vertebrae of the *metaphren* or thorax [231] and with the sternum ; these are called 'true' and 'perfect'. The remaining five are also joined as regards their posterior part to the spine, and of these the first three adhere anteriorly by their own cartilages to the true ribs, while the remaining two are separate ; these are called 'spurious', *nothae* or 'false' [232], but every one of the twelve is articulated to one of the twelve vertebrae.

Latin	**Greek**	HEBREW	*ARABIC*

(230) ṢÈLA'OTH = ribs, biblical ; singular, ṢELA'.

(231) The *metaphren*, spelt by Vesalius in Latin letters as though a current term, is a rare Homeric word equivalent to 'back', literally = behind the diaphragm. It is used in the much-read but spuriously Galenic *Introductio seu medicus*, but nowhere in such true Galenic writings as were available early in the sixteenth century. It was used by Valla (*c.* 1500), Benedictus (1502) and Berengar (1523) as equivalent to the part of the back or of the backbone between the scapulae, and more exactly by Charles Estienne (*c.* 1538) for the dorsal vertebrae. It gained currency for the part corresponding to the thoracic vertebrae, and is so used by Sylvius in his commentary on the *De ossibus* (Paris, 1543), and doubtless in the teaching which Vesalius received from him at Paris. The word became popular and survived into the nineteenth century even in the vernacular. It is never used now.

(232) The term 'false ribs', **pleurai nothai**, was used by Galen (*De ossibus*, Ch. XIV; Kühn, II, 765). It was adopted by Avicenna as *AL-KĀDHIBA* (I, 1; V, 1; 14) and translated by Gerard of Cremona as *falsae*. Avicenna (IV, 5; III; 7) introduced the term *AL-ṢĀDIQA = verae*. *Costae verae* and *costae falsae* are medieval commonplaces. Benedictus and Valla use the word *nothae*, which is perhaps the reason that Vesalius prints that Greek word in Latin letters, indicating that it had become current.

L. Very robust bones [233] which are joined to the *os sacrum* [234], GABH HA-'ERWAH *gafherua* [235]. The upper part is [*ossa*] **lagonōn** [236], *ilium ossa* [237], ALZARGÈPHĀ *alzargepha* [238], *anchae* [239] [N]. At the attachment of the femur [it is] **ischion** [240], [*os*] *coxendicis* [241], 'EṢEM HA-YAREKH *ezem haiarech* [242], pixis of the thigh [243], ALṬA'IGHA [244] *caph haiarech* [245], *althauorat* [246].

O. At the anterior part, where [the bones] are slender and perforated and mutually connected by synchondrosis, they are called [*ossa*] **hēbēs** [247], *pubis, pectinis* [248], *altaiga* [249], *penis*. The whole bone is called *os coxae* by Celsus [250], though by the author of the *Introductorium seu medicus* it is called **ischion** [251]. Some wrongly think these bones in the male are not united by cartilage at the pubes [252].

(233) The Vesalian nomenclature of the innominate bone is in confusion. This arises from differences between Galen and Avicenna. The former divides it into three, the latter into four bones, but neither along the lines of ossification. Galen, *De ossibus*, Ch. 20 (Kühn, II, 772) : " To the lateral parts of the sacrum are attached two bones, for which as a whole no name is available [that is, they are 'innominate ']. The upper flattened parts bear the name **lagonōn osta** [= ilia, see note 236]. The outer and lower parts behind the socket are **ischiōn osta**. The parts which are directed upwards and forwards, slender, fenestrated and united together at their ends, are **hēbēs osta** [see note 247]. Each **ischion** contains a very large socket connected by a very strong ligament to the head of the femur ('ligamentum teres ')". Avicenna (I, 1; V, 1; 25) : " Adjoining the sacrum are two bones which unite in the middle by a rigid articulation ('symphysis pubis '). They are the support for all the bones above and the fulcrum for all below. Each is divided into four parts. The upper part is called *AL-ḤARQAFA* or '*AZM AL-KHĀṢIRA* (= loin bone = iliac blade). That in front is called '*AZM AL-'ĀNA* (= pubic bone). That behind is called '*AZM AL-WARK* (= haunch bone = *os anchae* in Gerard). That below and tending inwards is called *ḤUQQ AL-FAKHIDH* (= box of the femur = *pixis coxae* in Gerard), because it contains a cavity in which the round head of the femur is enclosed ".

(234) On *os sacrum*, see note 324.

(235) GABH HA-'ERWAH = lit. back of the genitals. One of the few exclusively Hebrew terms that entered medieval medical terminology, though most Arabic ones entered through Hebrew and with a hebraized pronunciation. The transcription of the *Tabulae Sex* is in the current corrupt pronunciation. Properly the word can apply only to the os pubis. The transference of the name to the whole innominate bone is taken from Arabic usage in which *AL-'ĀNA* = she-ass (applied to the pubes, perhaps as a slang term, just as French

'la chatte ') also serves for both ideas. In the printed Hebrew Avicenna and in the *Fabrica* an attempt is made to distinguish the two by calling the innominate GABH HA-'ERWAH and the pubis 'ESEM [SHE-'AL] GABBE HA-'ERWAH = bone over the genitals. In the *Tabulae Sex* the Hebrew name of the os pubis proper is not given (but see below, note 244).

(236) **Lagōn** = flank. Galen speaks of **lagonōn osta**, *De anat. admin.* Bk. V, Ch. 6, and VIII, Ch. 10 (Kühn, II, 507, 702) = ilia.

(237) In classical Latin (Virgil, Horace, etc.) *ilia* (genitive *ilium*) is normally a third declension neuter plural = ' the flanks ', and is without a singular. We thus get a natural *os ilium* = lit. bone of the flanks, as correctly used by Vesalius. The modern usage, with *ilium* as a singular, is, however, defensible even on purist grounds.

(238) ALZARGÉPHĀ is a corruption of *AL-ḤARQAFA* = ilium. This word is variously presented in Hebrew manuscripts. The printed Hebrew Avicenna makes of it HARKABHAH = grafting (or were the editors thinking of the root RKB = to ride ?). This word is missing in the *Fabrica*, which has instead the second name of the bone, 'ESEM HA-KESEL = loin-bone = '*AZM AL-KHĀSIRA* = flank-bone, which is a translation of the Greek **lagonōn ostoun**. (See note 236.)

(239) *Anchae* is a difficult and interesting word. Its first known appearance is about 1070 in Constantine (*De communibus medico cognitu necessariis locis*, II, Ch. 8), where we read of a *pyxis anchae*. This is the acetabulum of our nomenclature. Constantine relied exclusively on Arabic sources, and the word may be Arabic. There is a word *ANQĀ*', used to designate any large bone containing marrow, *e.g.* brachium, tibia, and femur (Ibn Sīda, al-Mukhassas, I, 164). Thus *pyxis anchae* = *ḤUQQ AL-ANQĀ*' = box of the femur (cf. *ḤUQQ AL-FAKHIDH*, note 233 at end). If the word is really of Arabic origin, it is curious that it spread throughout the languages of Western Europe. The word was similarly used a century later by Gerard of Cremona (*c.* 1170) in his translation of Avicenna. French *hanche* appears as early as the twelfth century. In Provençal, Spanish, Italian, Portuguese and English there are variants—usually *hancha* in non-medical medieval Latin— from the thirteenth century onward. ' Haunch ' represents it in modern English. No plausible Romance or Teutonic etymology has been suggested and, on the whole, an Arabic origin conveyed by Hebrew physicians seems probable. Anatomically it is nearly always plural. Mundinus, together with several immediate predecessors of Vesalius (*e.g.*, Zerbi, Achillini) and contemporaries (*e.g.* Massa, Tagault, Joubert), use the word in a confused or at least an indefinite way.

(240) **Ischion** is Galen's term for the general area of bone that we now distinguish as a separate bone. The acetabulum he divides between ilium and ischium, excluding the pubis from its rightful share, *De ossibus*, Ch. XX (Kühn, II, 772).

(241) *Coxendix* is a rare classical term for hip-bone. Its employment by Vesalius to signify ischium is needlessly confusing.

(242) 'ESEM HA-YAREKH = hip-bone. YAREKH is biblical = thigh.

(243) *Pixis* or capsule of the coxa seems most applicable to the acetabulum itself, which is made into a separate bone by Avicenna. (See note 234.)

(244) AL-ṬA'IGHĀ. Here the text is in disorder. This word should stand before *altaiga*, two lines further. It is presumably an Arabic term for *os pubis*, but we cannot trace its source. Is it an arabicized form of one of the Romance developments of *theca* = pyxis (see W. Meyer-Lübke, ' Romanisches etymologisches Wörterbuch ', Heidelberg, 1935, no. 8699) ? If it is, then it must have been brought from Spain by Jewish doctors.

(245) KAPH HA-YAREKH = lit. spoon of the femur, is biblical = *ḤUQQ AL-FAKHIDH* = box of the femur.

(246) *Althavorat*. We have not found this word in Arabic medical literature. Perhaps it is *AL-FAWWĀRA* = groove or fold of the groin (cf. Ibn Sīda, al-Mukhassas, II, 42). *F* and *TH* frequently replace each other in Arabic. *Althavorat* would represent the plural *AL-FAWWĀRĀT*. It is an interesting problem how these non-technical Arabic words came into Western usage. Presumably Jewish doctors were the main source.

(247) **Hēbēs** [ostoun] = bone of the pubes. The word **hēbē** is connected with the verb **hēbaskein** = to come to puberty = Latin *pubescere*. **Hēbē** is used in a general sense for the pubes.

Latin	**Greek**	HEBREW	*ARABIC*

(248) Celsus calls the bone *os pectinis*. In Juvenal and Pliny *pecten* = pubic hair, thus the Celsan usage is by transference.

(249) *Altaiga*. See above, note 244.

(250) Celsus (VIII, Ch. 1, 23) calls the innominate bones *os coxarum*, treating them as one bone united at the symphysis pubis.

(251) The work to which Vesalius refers is the spurious *Introductio seu medicus* (Ch. VII, Kühn, XIV, 724), which appears as the work of Galen in the Giunta and most older editions. The caution of Vesalius in referring not to Galen but to 'the author of the *Introductorium*', is not based on his own judgment but on that of his teacher, Guenther, who produced the first Latin edition of it at Paris in 1528 (see dedication of that work). The book was commonly used in the schools by the time of the *Tabulae Sex*.

(252) Among these errant ones is Celsus. (See note 250.)

Q. *Tibia* (253) **knēmē** (254), 'ASMOTH HA-SHOQ *ahtzmoth hascok* (255), by which titles this whole part is named. QANEH GADHOL *cane gadol, canna major* (256), greater focile of the leg (257). The anterior part of this, fleshless and thin, is called *crea* (258), and it is to it that the femur is joined. Moreover the hollows of the tibia which receive the heads of the femur are easy to see.

(253) *Tibia* = flute. The instrument was originally formed from a long bone of an ungulate, especially from that which we now call *tibia*. Celsus (VIII, Ch. 1 and 11) uses the word in our modern anatomical sense. In medieval anatomies *tibia* = leg. Vesalius reverts here to Celsan usage.

(254) **Knēmē** = leg, but in Galen, *De ossibus*, Ch. XXII (Kühn, II, 774), it is the bone we call tibia. **Knēmē** survives in the modern anatomical term 'gastrocnemius'.

(255) 'ASMOTH HA-SHOQ = bones of the lower leg. SHOQ is biblical.

(256) *Canna*, classical Greek and Latin for reed, reed-pipe (our 'cane'). Biblical QANEH is also a reed. The word itself is one of the rare and interesting cases of a classical Graeco-Latin word of Semitic and, presumably, Hebrew origin, *cf.* Muss-Arnoldt, 'On Semitic words in Greek and Latin', *Transactions of the American Philological Association*, xxiii (1892), 108. The word is ultimately derived from Sumerian. QANEH GADHOL = large canna. The designation of the two leg-bones as 'canes', as that of the arm-bones as 'fire-sticks' (note 142), appears to have originated with the Arabic anatomical texts and to have passed from them into Hebrew and Latin. It occurs in Gerard's Latin Avicenna and became a medieval anatomical commonplace. In Arabic every bone that is hollow and contains marrow is called *QASABA* = cane (Ibn Sīda, al-Mukhassas, II, 164).

(257) For *focile* see note 142.

(258) *Crea* for *ocrea* = **okris** = greeve. The loss of the *o* is strange. The word is dropped in the *Fabrica* and has not been traced elsewhere in anatomical usage.

R. *Fibula* (259), *sura* (260), the lesser bone of the leg, **perōnē** (261), QANEH QAṬON *cave*(!) *katon* (262), *canna* and *arundo minor* (263). This bone is less thick than the tibia, nor is it so prolonged as to reach the knee itself. Both above and below it is united to the tibia by **synarthrosis**. The whole of this part is called *crus* by Celsus.

(259) *Fibula* = buckle (not its pin; see note 261), and has so persisted in various Romance languages. In post-classical times it acquired the sense of *spiculum* (Du Cange). *Fibula* entered anatomical nomenclature in the fourteenth century with Nicholas of Reggio (note 261).

(260) *Sura* = calf of leg, is used by Celsus as = fibula.

(261) **Perōnē** = pin, and especially the pin that forms part of a safety-pin or buckle, the instrument known in Latin as *fibula*. **Perōnē** is especially applicable to the splinter-like fibula of ungulates in which the shaft is always separate from the lower epiphysis. So Galen, *De usu partium*, I, Ch. 5 (Kühn, III, 9), denominates the fibula as **perōnē**. *Fibula* is used to translate **perōnē** by Nicholas of Reggio, who doubtless had in mind the two bars of the safety-pin.

(262) QANEH QATON = lit. small cane, opposed to the QANEH GADHOL = great cane, or tibia.

(263) *Arundo* is another word for reed or cane. Stephen of Pisa, translating Haly Abbas, uses it to render *QASABA*.

S, T. *Malleoli* [264], **sphyra** [265], ARKUBBOTH *arcuboth* [266], *claviculae* [267], are parts of the processes at the end of tibia and fibula.

(264) *Malleolus*, diminutive of *malleus* = hammer. No early application to anatomy has been found. The word seems to have entered the anatomical vocabulary in the sixteenth century. We have traced it first in Benedictus (1502), and frequently in the first Giunta Galen (see note 174). Massa (1536) uses the word *malleolus* as equivalent to the *malleus* of the middle ear.

(265) **Sphyra** = *malleus* = hammer, is used by Galen, *De ossibus*, Ch. 22 (Kühn, II, 774). Several glossaries from the seventh century onward give **sphyra** = *malleolus*. (Goetz, *Glossae*, II, 486–68 ; II, 126–38 ; III, 261–48.)

(266) ARKUBBOTH, plural of Mishnaic ARKUBBAH, can only mean knees. It is so used in the paragraph on the patella (see note 164). Rhazes and Haly Abbas have no name for the malleoli. Avicenna and his translators merely call them ' the two ends protruding from tibia and fibula '. The term ARKUBBOTH must have arisen in oral medical usage. The confusion may have been favoured by the existence of a word '*URQŪB* = a muscle at the back of the lower leg, connecting with the heel = gastrocnemius (Ibn Sīda, Mukhaṣṣaṣ, II, 53).

(267) Zerbi, Benedictus, and Berengar equate *claviculae* with *tarsi*. Thus Vesalius, in saying ' the claviculae are parts of the tibia and fibula ', is confusing the malleoli and tarsal bones.

V. Largest of all the bones of the foot, **kalkaneos** [268], **pterna** [269], *calcis os*, '**AQEBH** [270] *aekef* [271]. The posterior part of this passes far behind the line of the tibia.

(268) We cannot trace **kalkaneos** as a Greek word, and to write it in Greek letters may be a slip of Vesalius. *Calcaneum* is late Latin. It occurs for the first time in Ambrosius (*c*. 400) and about 600 in St. Isidore's *Etymologiae* (XI, 1, 114), which has the absurd derivation, ' The first part of the sole is the *calcis*, a name given it from the *callus* (=leathery skin), by which we trample (*calcamus*) the earth, and hence *calcaneus* '. The word is still current in Romance languages, *e.g*, Italian ' calcaneo '.

(269) **Pterna** is the normal term given to the calcaneum in the works of Galen.

(270) *Os calcis* is the name given to the calcaneum by Celsus, Book VIII, Ch. 1, 27.

(271) '**AQEBH** = *calcaneum*, is biblical. The transcription *aekef* suggests a popular corruption of the word to '**EQEBH**, *i.e.*, to the segolate pattern (see note 133) and oral transmission, in which final *v* became *f*. This would be the Ashkenazi pronunciation current among North Italian Jews. In the *Fabrica* the word is transcribed correctly as *haachev*.

Distich embracing the number of the bones :

Add to four times ten twice a hundred and six and thou wilt know Once and for all what a large number of bones it is of which thou art composed.

Latin   **Greek**   HEBREW   *ARABIC*

TABULA VI.

Skeleton depicted from behind.

The **raphai** [272], sutures [273, MĚHUBBARIM *mechubarim* [274], *complosa* [275], of the shape of the natural head which is formed in the likeness of an oblong sphere.

(272) **Raphai** = lit. stitchings, sewings, seams, is the term for the sutures of the skull used by the Greeks from Herodotus onward. It occurs frequently in Galenic works.

(273) The first element of *su-tura* is etymologically equivalent to the Teutonic root of English sew. *Sutura* is used for the sutures of the skull by Celsus, VIII, 1.

(274) MĚHUBBARIM would be Mishnaic Hebrew for ' connected ones '. It does not appear in the Hebrew Avicenna translations, which use HULYOTH—or SHĚLABBIM = links. Perhaps *mechubarim* translates *complosae* (see note 275). If so—that is, if the Hebrew derived from the medieval Latin term—we may assume that Vesalius received it via oral tradition. There is the possibility that the word is due to careless reading of some Hebrew medical text, such as the translation of the *Almansor* of Rhazes, made from the Latin (Bodleian MS Neubauer 2090), where it is said that the bones of the skull are MĚHUBBARIM = connected with each other. The Arabic Avicenna calls the sutures *DURŪZ* = seams, whence such medieval corruptions as *direzan, adorez, adoren*, etc., or SHU'ŪN, whence medieval *asoan* (see note 195). The *Fabrica* drops *mechubarim* and uses SHĚLABBIM *scelavim* instead.

(275) *Complosa* is a barbaric word of medieval origin. In Stephen of Pisa's Latin translation of Haly Abbas (*Theorice*, II, Ch. 3, and elsewhere) *complosiones* = sutures. Examination of the text shows that this is a misreading of *complōnes*, which is a contracted form of *complexiones* = interweavings = complex sutures. In other medieval works are forms such as *complosus* and *complosio*, standing for complexus, complexio. (See Ducange and also J. H. Baxter and C. Johnson's *Mediaeval Latin Word-list*.)

A. Across the *sinciput* [276] : **stephaniaia** [277], *coronalis*, as place of the crown [278], KITHRI *cethari* [279]. It appears in the other figures [but not in this].

(276) See note 197.

(277) **Stephaniaia** is found for the first time in Rufus, *De nom. part.* (Daremb. and Ruelle, p. 151). It is the term used by Galen.

(278) *Coronalis* is found first in Stephen of Pisa's Latin translation of Haly Abbas as *coronatum.* The form *coronalis* is due to Gerard.

(279) KITHRI, from biblical KETHER = crown. The transcription *cethari* suggests a popular mispronunciation. Moreover, it is a translation from the Latin *coronalis* rather than from the Arabic, for the original Greek **stephaniaia** = garland-suture is rightly rendered by Avicenna *IKLĪLĪ* and by the translators Meathi, Lorci and the Hebrew printed Avicenna KĚLILI, which is also given in the *Fabrica*.

B. Across the *occiput* [280] : from the form of the letter Λ, **labdoeidēs** [281] HA-ṬETH *hateth* [282], *laude* [283].

(280) See note 209.

(281) **Labdoeidēs** is in Galen, *De ossibus*, Ch. 1 (Kühn, II, 740).

(282) HA-ṬETH = lit. the letter ט. The Arabs called it *AL-LĀMĪ*, their Hebrew translators ḤA-LAMDI, always pointing out that it is the Greek Λ that is meant, since neither Hebrew ל nor Arabic ل resemble it in form. The letter ט is in medieval MSS. pointed at the bottom, and therefore an apt substitution. Here again Vesalius drew on the current jargon of his Jewish contemporaries. In the *Fabrica* only the name LAMDI appears.

(283) *Laude* is a barbaric corruption through *lavde, labde*, **labda,** and is a medieval anatomical commonplace.

C. From the middle of the posterior suture, through the vertex anteriorly : **oboliaia** (284), as being arrow-like, HA-ḤEṢ *hachets* (285), *sagitalis* (286), *nervalis* (287). Sometimes it ends between the eyebrows, cleaving the frontal bone itself, but it is untrue that this is so in all women and in no men (288).

(284) **Oboliaia** would mean worth a farthing (an obol) ! It is a mistake for **obeliaia** = like a skewer. It is from the pseudo-Galenic *Introductio seu medicus*. Oddly enough the same mistake occurs in Guenther's editio princeps (Paris, 1528), in Kühn (XIV, 720) and in the Giunta editions ! The term does not occur in genuine Galenic works. There one finds **hē kata to mēkos eutheia** = the straight one through the length, *De ossibus*, Ch. 1 (Kühn, II, 742). Rufus (p. 151) has **epizeugnyousa**, 'the uniting one '.

(285) **HA-ḤEṢ** = lit. the arrow. In the Hebrew translations of Avicenna and in the *Fabrica* called **HA-ḤIṢṢI** = *AL-SAHMĪ* = the arrow-like. The form found here again gives the impression of being subject to the vicissitudes of oral transmission.

(286) The arrow simile comes from Arabic literature. It is first found in Haly Abbas (died 994), who uses it along with the Galenic 'straight suture ' = *AL-MUSTAQĪM*. Stephen of Pisa introduced the simile to the Latins in 1127. The actual form *sagittalis* arrived with Gerard, *c.* 1170.

(287) *Nervalis* cannot be traced. *Nervus*, meaning both nerve and tendon, came to signify bowstring (made from tendon). Arabic anatomical MSS usually have a diagram of the sutures likened to a bow. It almost seems as if *nervalis* had been at first a designation for *labda*, the string of the bow. The name *arcualis* was sometimes employed for the coronal suture.

(288) Aristotle, *Hist. anim.* 516, a. 16.

These three sutures vary in the three non-natural forms of the head by loss or retention of prominences (289).

(289) The discussion of head-forms begins in the Hippocratic writings (*De capitis vulneribus*, Ch. 1, and elsewhere). The theory of three possible non-natural forms is propounded by Galen, *De usu partium*, I, Ch. 17 (Kühn, III, 752 *f*). He denies the possibility of a fourth non-natural form. One would like to regard these passages as beginning craniometry, but as they stand they have no relation to the facts. More intelligible are the famous skull-figures in the *Fabrica* (p. 17).

D. At the temples are two **krotaphiai** (290), *temporales, corticales* (291), **lepidoeides** (292), *squammaeformes* (293), SHĔḤUSI *schechusi* (294). They are to be regarded as commissures rather than sutures.

(290) **Krotaphoi,** a Galenic term for ' temples '.

(291) *Corticalis* = bark-like [suture], from *cortex*. It was first used by Gerard to render *AL-QISHRIYYĀNI*. That word, taken by Avicenna from Ḥubaish, was employed by the latter to translate **lepidoeides** (Simon, ' Sieben Buecher, etc.', I, 13). Gerard somewhat misunderstood it, believing it to be from *QISHR* = bark, instead of from *QISHRA* = scale. (See note 292.)

(292) **Lepidoeidēs** = lit. scale-like, referring to the form of the upper part of the temporal bone where it is thin and greatly overlaps the parietal. It is a Galenic commonplace.

(293) *Squammaeformes* translates **lepidoeides.** It has no classical foundation. We have not been able to trace it further.

(294) SHĔḤUSI should be HA-SHĔḤUSIYYIM = lit. the cartilaginous [sutures]. SHĔḤUS is used by Lorci and other translators for cartilage. Its etymology is unknown. The Mishnaic word for cartilage is HASHUS, החסום, which was perhaps misread as החשום, HA-SHEHUS, an easy confusion with Rabbinic script. Why cartilaginous ? Th[e]

Latin **Greek** HEBREW *ARABIC*

seems to be a whole chain of confusions. *AL-QISHRIYYĀNI* (see note 291) was misinterpreted as ' parchment-like ', and accordingly by all Hebrew Avicenna translators rendered as HA-QĔLAPHIYYIM. But QELAPHI was used by some, *e.g.* Graciano, for ' cartilaginous ', and oral tradition, taking the word here in this sense, substituted for it its synonym SHĔḤUSI.

The remaining [suturae of the skull] are blendings rather than sutures ; one extending from the **labdoeidēs** (' occipito-temporal articulation '); another, extending from either [end of] the coronalis to the *os cuneiforme* (' articulation with greater wing of sphenoid '); one from the end of this through the midst of the eye-sockets to the root of the nose (' articulation with lesser wing of sphenoid '); one from the temporal fossae through the lower border of the orbit (' maxillo-ethmoid and maxillo-palatine articulations '); from which again three small [sutures extend] in the orbit defining the three bones of the nose [namely, ' nasal, frontal process of maxilla, and nasal part of frontal ']. Those described by Galen as extending from the canine teeth on to the outer part are invisible, to me at least. One [suture] which cuts transversely across the palate (' maxillo-palatine articulation ') and another which extends along the middle line of the palate to the openings of the nares (' intermaxillary and interpalatine articulation ') cannot be represented here.

(295) A passage in Galen's *De ossibus*, Ch. 3 (Kühn, II, 749) describes a suture that ' goes right through between the canines and incisors '. Avicenna follows Galen, schematising the lines of division. In one MS of the *Canon* these are even expressed in a geometric diagram (reproduced in de Koning, p. 463). The suture between canines and incisors, invisible to Vesalius, represents the demarcation between premaxillae and maxillae, which disappears in man in the third month of foetal life (except for a faint trace on the palate). It is well marked on the front of the skull of the Macaca monkey, which Galen was describing. Sylvius, in his *Commentarius* on *De ossibus*, sees it clearly in man ! This passage in the *Tabulae*, with several others which indicate the doubts of Vesalius as to Galen's reliability, foreshadow the well-known controversy between Sylvius and his pupil. Nevertheless, Vesalius was not the only doubter, and he, moreover, often accepted Galen when Galen was wrong.

E. The *ossa lapidosa* (296), *petrosa* (297), **lithoeidē** (298) have no proper limits [*i.e.* no sutures] but that portion which is near the base of the skull, through which the *medulla spinalis* emerges, is properly so named. Some suggest that the [whole] bone of the temple should be known by this name (299).

(296) *Lapidosa* is the term used by Sylvius in his *Commentarius in librum Galeni de ossibus*. It was perhaps taken from Stephen of Pisa's Latin translation of Haly Abbas, where it appears as *lapidea*. *Lapidosa* was doubtless a term of the Paris school, but gained no currency elsewhere.

(297) *Petrosa* is the normal medieval term for this process. It was introduced by Gerard in his Avicenna translation, is used by Mundinus and others, and has survived into modern nomenclature.

(298) **Lithoeidē** is used by Galen, *De ossibus*, Ch. 1 (Kühn, II, 745). It is given in Guenther's version of the pseudo-Galenic *Introductio seu medicus* (Paris, 1528). Note that the *Tabulae Sex* gives here no Hebrew word. The *Fabrica* has HA-ABHNIYYIM = the stone-like, from the printed Hebrew Avicenna. Again it is clear that the Hebrew adviser of Vesalius for the *Tabulae Sex* relied on MSS, while the adviser for the *Fabrica* used printed sources.

(299) The word ' some ' may refer to the *Introductio seu medicus*, Ch. 12 (Kühn, XIV, 721).

F. *Scapula* [300], **omon** [301], **ōmoplatē** [302], sometimes called by the Latins *humerus* [303], *scoptulum opertum* [304], *scoptula*, KATHEPH [305] *cathep*, *spatula* [306].

(300) In classical usage *scapulae* occurs only in plural and means ' back '. Thus in Celsus the shoulder-blade is *os latum scapularum* = broad bone of the back. The singular arrives with the Vulgate, that is about 400 A.D. In the Middle Ages the bone was usually called *spatula* (see note 306). Valla, *De corp. part.* (1528), restores the classical usage. Berengar (1523) gives *scapulae* = *spatulae* = *scapillum*. Since Guenther, *Inst. anat.* (1536), uses only **ōmoplatē** (see note (302)), it may be that Vesalius is here using *scapula* in the modern sense for the first time.

(301) **Ōmos** (not **ōmon**) is the normal classical term for ' shoulder '. **Ōmos** is the etymological equivalent of *hum-erus*. But **ōmos** can hardly mean shoulder-blade, which is represented by **ōmoplatē**.

(302) **Ōmoplatē** = lit. the broad of the shoulder = *scapulae*, the normal classical form for shoulder-blade, and preserved in modern French in original form. It entered French with Charles Estienne's translation of his own work in 1546.

(303) On *humerus*, see note 111.

(304) Celsus calls the shoulder-blade *scutulum opertum* = lit. little covering shield. *Scoptulum* is a barbarism which we have traced to Valla (*c.* 1500). It is, perhaps, formed with the memory of *scutulum*, *scopulus* (= a projecting point of rock, crag) and *spatula* (see note 306).

(305) KATHEPH = shoulder-blade, biblical.

(306) *Spatula* is a common medieval term. It is not, as might be thought, a barbarism for *scapula*, but is from **spathē** = blade of a broad-sword. Isidore in his *Etymologiae*, XVIII, vi, 4 (*c.* 600) and Constantine in his *Pantegni*, Ch. 6 (*c.* 1070), use the word in the anatomical sense of *scapula*.

G. Cervix or neck of the scapula surrounding the acetabulum [307].

(307) *Acetabulum* = originally a small vessel for vinegar (*acetum*). It is used for the socket of the hip-bone by Pliny, than whom no secular writer was more read in the Dark Ages. In Pliny's sense it passed into medieval and modern anatomical usage. Valla (*De corporis partibus*, edition of 1536 printed with Guenther's *Inst. anat.*) gives the form *acetabulum humeri*.

Rhachēs [308], *spina dorsi*, SHEDHRAH *schedra* [309], consists of twenty-four [310] **spondyloi** [311], *vertebrae* [312], ḤULYOTH *chulioth* [313] : seven of the neck, **trachēlos** [314], MEGHARON *megaron* [315]; twelve of the back, **notiaion** [316], *metaphrenum* [317], *thorax*, HA-GABH *hagab* [318] ; five of the loins, **osphys** [319], *lumbi*, HE-'AṢEH *heatze* [320], *renes*, *alchatin* [321] ; hitherto I have seen the eleventh, but not the tenth dorsal vertebra to be reduced, and the posterior processes called *spinae* and *simenia* [322] of the vertebrae after the tenth to be reduced or rather not to increase. But only the twelfth, as the Arabs agree, have I seen to be without a transverse process [323].

(308) **Rhachis** = backbone, classical and used by Galen. The form **rachēs** suggests late vulgar Greek.

(309) SHEDHRAH = backbone, Mishnaic Hebrew.

(310) The elements of sacrum and coccyx were not reckoned as vertebrae by the earlier anatomists.

(311) **Spondylos**, more correctly **sphondylos** = a vertebra. Galen uses the form **spondylos**.

(312) *Vertebra* is in classical Latin, and even in Celsus, a joint in general, gradually passing into its specialised sense in late Latin.

(313) ḤULYOTH, Mishnaic Hebrew, sing. ḤULYAH = lit. link of a chain, but occasionally = a vertebra, *e.g.*, Bab. Talmud *Berakhoth*, folio 28 *b*.

Latin	**Greek**	HEBREW	*ARABIC*

(314) **Trachēlos** = neck region.

(315) MEGHARON for [ḤULYOTH] HA-GARON = [vertebrae] of the neck (see note 313). GARON
is biblical. The error reprcduced by the transcription, ME- for HA-, is strange, and possibly
assimilated to a Greek form.

(316) **Notiaion** for nōtiaion = back. Again a vulgar Greek source is suggested (see note 308).
The adjective also demands a noun, *e.g.* **nōtiaia arthra** = the thoracic vertebrae.

(317) *Metaphrenum*, see note 231.

(318) [ḤULYOTH] HA-GABH = vertebrae of the back. For ḤULYOTH see note 313. GABH=
back, biblical. Among Arabic anatomical writers, the nomenclature of the upper vertebrae
was confused. Rhazes used *KHARAZ AL-ZAHR* = vertebrae of the back, for the thoracic
together with the lumbar vertebrae. Haly Abbas used the same term for the thoracic
vertebrae alone. Avicenna dropped the term and called the thoracic vertebrae *KHARAZ
AL-ṢADR* = vertebrae of the breast. In the Hebrew translators the confusion is further
increased by combining the various nomenclatures. The Hebrew term given here might go
back to a Hebrew translation of Haly Abbas. How popular the term was is proved
by its occurrence in the Hebrew translation of the *Almansor* of Rhazes, made from the
Latin (Bodleian MS Neubauer 2090), where it translates *spondiles pectoris* in the
Latin text. The *Fabrica* has ḤULYOTH HE-ḤAZEH = vertebrae of the chest, the
Avicennan term, but it is not taken from the Hebrew printed Avicenna. There another
combination is found—ḤULYOTH HA-SHEDHRAH WĔ-HE-ḤAZEH = vertebrae of
the spine (!) and the breast. (For SHEDHRAH see note 309.)

(319) **Osphys** = loins. Used by Galen.

(320) [ḤULYOTH] HE-'AṢEH = vertebrae of the coccyx. The word has wrongly changed places
with HA-QAṬON = the lumbar region (see note 332). 'AṢEH is biblical (Leviticus, iii, 9).

(321) *Alchatin*, a very familiar term of medieval anatomy, represents *AL-QAṬN* and derives from
Gerard's Avicenna. (See note 332.)

(322) *Simenia* is one of the many corruptions of *SANĀSIN*, plural of *SINSIN* = thorn = spine of
vertebrae. Gerard uses the form *senasen*. The form *simemia* is found in the thirteenth
century pseudo-Galenic *De anatomia vivorum*.

(323) The passage describes the gradual reduction of both transverse processes and spines in descending
the thoracic vertebrae. Normally the transverse processes of the eleventh vertebra are
most reduced. In placing the minimum at the twelfth, Vesalius is, as he says, in agreement
with the Arabic anatomists, as represented by Avicenna, I, 1 ; V, 1 ; 7 (De Koning, p. 482).

> H. *Os* **hieron**, *sacrum* (324), **platy**, *latum, amplum* (325), AL'AGHAZ *alagas* (326),
> *osanium* (327) *alchavis* (328), *hagit* (329). Galen in his book *De ossibus*
> put out that the sacrum is formed of three bones, in his *De usu partium*
> of four bones (330).

(324) The term *os sacrum* = sacred bone, translates **ostoun hieron**, in which the sense of **hieron** has
been said not to be sacred but big or solid. This statement, though often made since the
sixteenth century, has little or no classical backing, and the origin of the name remains a
mystery. However, **ostoun hieron** is the ordinary Galenic term and is of Hippocratic origin,
De articulis, 45. The first occurrence of its Latin form, *os sacrum*, is in the fifth century
with the Numidian physician Coelius Aurelianus. It is a medieval commonplace.

(325) None of these three terms for the sacrum, all implying breadth, gained any anatomical currency.
Platy is used by Galen, *De anat. admin.* VI, Ch. 14 (Kühn, II, 585) ; *latum* is used by
Guenther in his translation of the same work in the Giunta edition ; *amplum* we have not
traced.

(326) AL'AGHAZ = *AL-'IJZ, AL-'UJZ, AL-'AJUZ* = the buttocks. The Arabic term is transliterated by the Hebrew translators of Avicenna, Meathi and Lorci, so that it was probably
current in Hebrew medical parlance. This is also suggested by its assimilation to the
segolate pattern (see note 133). Graciano translates it in one place as HA-ZAQEN = the
old man, as if it was *AL-'AJŪZ*, and in another as HA-'EṢEM SHE-'AL HA-AMMAH=
the bone above the penis. The printed Hebrew Avicenna hebraizes into 'EṢEM PI HA-
ṬABBA'ATH=anus-bone, which appears in the *Fabrica*.

(327) *Osanium* we have not specifically traced, but it is surely *os ani* run into one, misunderstood and with *-um* added to give a Latin form. For similar Hebrew formation see note 326.

(328) Gerard of Cremona writes in his *Expositio nominum Arabicorum* that *Alharis sunt ossa lata quae sunt sub renibus parum* = ' Alharis are broad bones which are a little below the kidneys ' = innominates. But in his Latin version of Avicenna's *Canon* Gerard renders *AL-'IJZ* as *sacrum*. In medieval anatomical texts the word takes such forms as *Alanis, Halaris, Alhovius*, etc. It is derived from *AL-'IJZ* = the buttocks.

(329) *Hagit* we have not traced. It may have originated through the misreading of some such form as *hagiz* in a Latin text, or it may be derived from *'AJB* = part of the os sacrum or coccyx adjoining the innominate bone, by supplying the wrong diacritic points (عجِز instead of عجْب).

(330) Another passage which expresses doubt about Galen's reliability (see note 295). In *De ossibus*, Ch. 12 (Kühn, II, 762), Galen is sure that there are three vertebrae in the sacrum. In *De usu part.* XII, Ch. 12 (Kühn, IV, 50), he says : ' As to the *os sacrum* I shall show that it is better to treat it as consisting of four vertebrae '. In fact, it consists normally of five vertebrae, though human sacra of four and of six are not extremely rare. In Macaca it consists normally of three and there are seven lumbar vertebrae. In certain allied forms the pelvis consists of four vertebrae and there are but six lumbars. Galen had perhaps skeletons of two species of ape. In the skeleton of the *Tabulae Sex* the sacrum consists of four bones. In the *Fabrica*, p. 79, Vesalius redresses the balance by figuring a sacrum of six vertebrae. It is followed by the sacrum of an ape with the correct number of three.

I. Vertebrae of the *coccyx* [331], HA-QATON *hakaton* [332], *cauda* [333], *alhosos* [334], is made up of three bones, according to Galen in his books *De administrationibus anatomicis* and *De ossibus*, but he does not specify in his *De usu partium* [335]. Of a truth, the descriptions of Galen, Prince of Medicine, are not only inconsistent but we have observed that they do not suit these two bones at all, for we have found nine bones and fewer foramina than he describes [336].

(331) **Kokkyx** is Greek for cuckoo. How it came to be applied to a bone is a mystery, but it is so used by Galen. A parallel might perhaps be found in the Arabic *QATĀT* = sand-grouse, but also = croup, buttocks. Both terms seem to have originated in slang. For Pollux the sacrum is one of the **kokkyges.**

(332) **HA-QATON** = loins = *AL-QATN*, misplaced in text (see note 320). The form is due to confusion with QATON = small, and must have arisen in the oral tradition, as it is not found in the MSS. The Bodleian MS of Graciano vocalizes distinctly HA-QOTN. Against all earlier translators, the printed Hebrew Avicenna uses a Hebrew term, MOTHNAYIM, which appears in the *Fabrica*.

(333) *Cauda* occurs in Gerard's translation of the *Almansor* of Rhazes as a gloss to *alhosos*. It may translate the other Arabic term for the coccyx, *ASL AL-DHANAB* = root of the tail.

(334) *Alhosos* = *AL-'US'US* = the coccyx, also *AL-'USUS*, which is probably connected with the biblical 'ASEH. It may be derived from a root meaning ' to be hard '. According to Arab popular belief, the top part of the coccyx (*'AJB*, see note 329) never disintegrates. The form *alhosos* occurs in Gerard's translation of the *Almansor* of Rhazes.

(335) The reference to *De anat. admin.* is not to be found in Kühn or the Giunta editions. It may be that Vesalius is referring to a commentary. The reference to *De ossibus* is Ch. 12 (Kühn, II, 762). The place in *De usu part.* where the specification is wanting is XII, 2 (Kühn, IV, 50).

(336) This passage makes quite clear that in 1538 Vesalius was already definite as to the fallibility of Galen. As regards the nine bones, he was obviously treating coccyx and sacrum as one, under the influence of Pollux (see note 331).

Latin	**Greek**	**HEBREW**	*ARABIC*

K. **Myron** [337], *foemur* [338], *foemen* [339], PAḤADH HA-YAREKH *pachad haiarech* [340], the bone of the thigh. It is the greatest of all the bones of the body.

(337) **Mēros** = thigh, becomes thigh-bone in both Hippocrates and Galen. The form **myron** suggests a vulgar or oral tradition (*cf.* note 196).

(338) *Femur* is correct, not *foemur*. In classical usage = thigh.

(339) *Foemen* is a form imagined by grammarians and mentioned only by Priscian and Servius Vergilius. It is here for display.

(340) PAḤADH HA-YAREKH = Arabic *FAKHIDH* with Hebrew YAREKH, both meaning femur. The combination of the two appears in Meathi and Lorci. It may have arisen because the Hebrew YAREKH is too ambiguous, meaning 'hip', 'thigh', and 'pubes'. Graciano, and the printed Hebrew Avicenna, have only HA-YAREKH, but the *Fabrica* keeps to PAḤADH HA-YAREKH. In Job xl, 17, we find PAḤĂDHAW, probably meaning 'his thighs'. Perhaps the translators took the word from there because of its similarity with Arabic.

L. Upper head [of femur]. M. Neck PP [NN by error]. Lower heads with which the tibia articulates.

N. [O by error]. Great **trochantēr** [341], which they also call **gloutos** [342], *malum granatum testiculorum* [343].

O. [P by error.] Lesser and inner **trochantēr**.

(341) **Trochantēr** [trechein = to run], used by Galen, *De ossibus*, Ch. 2 (Kühn, II, 773).

(342) **Gloutos** is occasionally used by Galen, *De ossibus*, Ch. 21, for great trochanter.

(343) On *malum granatum*, see note 128.

Printed at Venice by B[ernardo] Vitali of Venice at cost of Stephen Calcar [344]. For sale [*prostrant* for *prostant*] in the shop of Signor Bernardo. Anno 1538.

(344) It seems remarkable that Calcar should have paid the cost of printing. We suggest that the meaning is that the whole issue was made over by Vesalius to Calcar as though the latter was responsible for the cost and that Calcar was to be recompensed by the profits of sale.

Warning.—By Decree of Pope Paul III and of his sublime Imperial Majesty, and of the Most Illustrious Senate of Venice : Let no one either print or retail or hawk separately these plates of Andreas Vesalius of Brussels under the heavy penalties set forth in the copyright.

IX. General Index.

Latin and Latino-Arabic words, modern anatomical terms,
proper names, and subjects.

*Where a word appears both in the translation and the notes attached to it, it is
given under the page where it appears in the translation.*

X. Greek Index.

A.

ἀγκυροειδής, 21.
ἀδενῶδες σῶμα, lxi.
ἀείρειν, 13.
ἄζυγος, 9.
ἀκρομφάλιον, ἀκρόμφαλον, lxx.
ἀκρώμιον, lxxxiv, 20.
ἀμυγδάλη, lxx.
ἀναστόμωσις, lxx.
ἀντιάς, lxx.
ἀορτή, lxxxiii, 13.
ἀπευθυσμένον ἔντερον, 11.
ἀπόπληκτος, 6.
ἀραιός, 5.
ἀρτηρία μεγάλη, 13.
ἀρτηρία ὀρθή, 13.
ἀστράγαλος, 26.

B.

βασιλικὴ φλέψ, 7.
βελονοειδής, 30.
βραχίων, 23.
βρέγμα, 30.

Γ.

γλουτός, 44.

Δ.

δάκτυλοι, 26.
δικτυοειδὲς πλέγμα, 14.
δωδεκαδάκτυλον, 4.

E.

ἔκφυσις, 4.
ἑξάγωνον, 27.
ἐπιγονατίς, 26.
ἐπιζευγνύουσα (ῥαφή), 39.

Z.

ζυγωειδῆ (ὀστᾶ), 31.
ζυγώματα, lxxxiv, 31.
ζῷον, xlviii.

H.

ἥβη, 35.
ἥβης (ὀστᾶ), 34.
ἧπαρ, 6.

Θ.

θαῦμα μέγιστον, 14.
θυμός, 8.

I.

ἱερόν (ὀστοῦν), 42.
ἰνίον, 30, 31.
ἵξις, 1.
ἰσχίον, 34.

K.

καίρια, 19.
καλκάνεος, 37.
καλκοειδῆ (ὀστᾶ), 27.
καρπός, 25.
καρωτίδες φλέβες, lxxxiv, 15.
τὸ κατὰ μέτωπον ὀστοῦν, 30.
ἡ κατὰ τὸ μῆκος εὐθεῖα (ῥαφή).
  39.
κατακλεῖδα, 20.
κενός, 22.
κερκίς, 25.
κεφαλικὴ φλέψ, 7.
κλείς, 20.
κνήμη, 36.
κοίλη (φλέψ), 6.
κοιλία, 6.
κόκκυξ, 43.
κορακοειδής, 20, 21.
κόραξ, 21.
κορυφή, 30.
κορῶνον (κορώνη), 33.
κρανίον, 29.
κροτάφιαι (ῥαφαί), 39.
κρόταφοι, 30.
κτείς, 26.
κτένιον, lxxxiv, 26.
κυβοειδής, 27.
κυνοδόντες, 19.
κωρώνη, 21.

Λ.

λαβδοειδής (ῥαφή), 38.
λαγών, 34.
λεπιδοειδής (ῥαφή), 39.
λιθοειδές (ὀστοῦν), 40.

M.

μαστοειδής, 19, 30.
μασχαλιαία (φλέψ), 7.
μεσάραιον, 5.
μεσεντέριον, 5.
μέση (φλέψ), 8.
μετακάρπιον, 26.
μετάφρενον, 34, 41.
μέτωπον, 30.
μηρός, 44.
μύλη, 26.
μυλίται (ὀδόντες), (in text μύλεται), 19.
μύρον, 44.

N.

νόθος, 34.
νωτιαῖον, 41.
νῶτος, lxxx.

Ξ.

ξιφοειδής, 23.

O.

ὀβελιαία (ῥάφη), 39.
ὀβολιαία (ῥάφη), 39.
ὀδόντες, 19.
ὀκρίς, 36.
ὀστέον στέρνου, 22.
ὀσφῦς, 41.

Π.

πάγκρεας, 4.
παραστάτης ἀδενώδης, lxi.
πεδίον, 28.
περιστεφανοῦσα (φλέψ), 10.
περόνη, 36.
πῆχυς, 24.
πλατύ (ὀστοῦν), 42.
πλάτυσμα μυοειδές, xxxiv.
πλευρά, 33.
πολύμορφον (ὀστοῦν), 31.
πτέρνα, 37.
πύλαι, 3.
πυλωρός, 4.

XI. Arabic Index

XII. Hebrew Index.

* The explanation on p. 20 is to be rectified. The Hebrew is a translation of *AL-ṬAWĀḤĪN* =millstones=*mylai*, the term employed by Haly Abbas (De Koning, p. 120), or of the popular form *AL-ṬAWĀḤĪN*=the grinding women, which is given by the Arabic dictionaries, and has also penetrated into Syriac as ṬAḤḤĀNĀTHĀ (the Syriac word is known only from the dictionaries and does not occur in written Syriac).

I

* To the comments on p. 29 add that the Arabic *JUMJUMA* = skull is sometimes used for a wooden drinking-bowl, *cf.* Peñuela, ' Die Goldene ', p. 91.

XIII. Facsimile of the 'Tabulae'.

⊱PRAESTANTISSIMO CLARISSIMOQVE VIRO DOMINO

D. NARCISSO PARTHENOPEO, CAESARIAE MAIESTATIS MEDICO PRIMARIO.

Domino suo et patrono, Andreas VVesalius Bruxellensis S.D.

NON ita pridem, Narcisse doctissime, quum Patauii ad medicinæ chirurgicæ lectionem delectus, inflammationis curationem pertractarem, diui Hippocratis & Galeni de reuulsione ac deriuatione sententiam explicaturus, uenas obiter in charta delineaui, ita ratus quid per κατ᾽ἴξιν Hippocrates intellexisset facile posse demonstrari. Nosti namque quantum hac tempestate, ea dictio distentionum atque contentionum, etiam inter eruditos, de uena secanda concitauerit, dum alii fibrarum consensum ac rectitudinem, alii aliud nescio quid, indicasse Hippocratem affirmant. Verum illa uenarum delineatio tantopere medicinæ professoribus studiosisque omnibus arrisit, ut arteriarum quoque & neruorum descriptionem, à me obnixè contenderent. Quia uerò ad meam pertinebat professionem Anatomes administratio, ipsis deesse non debui, potissimum quum scirem eiusmodi lineamenta, his qui secanti adfuissent, non mediocre commodum allatura. Aliàs siquidem aut partium corporis, aut simplicium pharmacorum cognitionem ex solis picturis, seu formulis uelle assequi, ut arduum, sic quoque uanum ac impossibile omnino arbitror, sed ad memoriam rerum confirmandam apprimè conducere, nemo negauerit. Cæterum cum plurimi hæc frustra imitari conarentur, rem prælo comisi, atque illis tabellis, alias adiunximus, quibus meum σκέλετον nuper in studiosorum gratiam construxi. Ioannes Stephanus, insignis nostri seculi pictor, tribus partibus appositissimè expressit, magno sanè usu eorum, qui non modo honestum, aut pulchrum, sed etiam utilè ac necessarium iudicant summi opificis solertiam artificiúmque contemplari, & domicilium illud animæ (ut Plato ait) introspicere. Præterea singulis partibus, quàquam id in præsenti negocio non admodum ex sententia confici potuit, sua nomina ascripsimus, barbaris, quæ etiam peritiores in plurimorum libris subinde remorari solent, minimè prætermissis. Quòd autem ad rei ueritatem attinet, nullum hic apicem ductum puta, quem Patauini studiosi in huius anni confectione, à me demonstratum non attestabuntur: ut interim sileam de Parisinis præceptoribus meis longe doctissimis & Louaniensibus medicis, apud quos non semel Anatomen publicè administraui. Porrò ut nouus hic noster conatus, alicuius patrocinio commendatior in lucem auspicato prodeat & ancipitem iudiciorum aleam securius experiatur, celebritati illum nominis tui nuncupare uisum est: partim quòd præter incomparabilem uariarum linguarum cognitionem, eximiam quandam singularémque Anatomes, sicuti etiam medicinæ & philosophiæ, scientiam adeptus is: adeò ut meritò apud nationes omnes, tanquam præcipuum medicorum & literatorum hominum, decus ac ornamentum ab 'eruditissimis quibúsque prædiceris: deinde quòd inter clarissimos uiros ea polleas animi prudentia, integritate, mira erga omnes naturæ mansuetudine & gratia: ut CAROLVS QVINTVS Inuictissimus Romanorum imperator semper Augustus, acerrimus ingeniorum æstimator, non suæ dumtaxat sanitatis tuendæ, aut amissæ recuperandæ præcipuum tibi locum concrediderit: uerum etiam te uniuersis Regni Hispaniæ ac Neopolitani medicis pharmacopolarumque officinis, in florente etiánum ætate tua, ceu fidissimum censorem præfecerit, compluribúsque honoribus & muneribus, inter ot præclaros alioquin uiros, amplissimè illustrauerit. Suscipe itaque Vir ornatissime, hoc chartaceum munusculum, ea humanitate, qua me quondam excepisti, dum non exiguis beneuolentiæ signis, animum erga me tuum peculiariter declarasti: quod si gratum tibi ac studiosis fore intellexero, aliquando maiora adiiciam. Vale Patauii Calé.Apri.An.salutis.M.D.XXXVIII.

⊱IECVR SANGVIFICATIONIS

OFFICINA, PER VENAM PORTAM, QVAE GRAECIS ϛελεχιαία, Arabibus vero ﻣﺴﺎﺭﻳﻘﺎ varidhascoer appellatur, ex ventriculo et intestinis chylum transumit, ac in lienem melancholicum succum expurgat.

GENERATIONIS ORGA-

NA, SVPERIVS VIRI, INFERIVS MVLIERIS.

Tertia figura semen deferentium vasorum implantationem refert.

A Cauum, seu simum iecoris.
B Vena porta, iecoris manus.
C Ramati in flauæ bilis vesiculã.
D Ad pancreas et ecphysim, seu duodenum intestinum.
E Ad dextrum gibbi ventriculi.
F Ad dextrum fundi ventriculi et superiorem omenti membranam.
G Portæ bifurcatio maxima.
H Per omenti inferiorem membranam et pancreas delata, varie diffunditur.
I In omenti membranam inferiorem, parte dextra.
K Per ventriculi cauum, eius os tandem numerosis propaginibus amplectens.
L In membranam omenti inferiorem parte media, quæ primum in duas, deinde in plurimes exiguas venulas diuaricatur.
M Multifariam diuisa, per rectam lineam liens simo implantatur: hac fæculentus sanguis in lienem transmittitur.
N Vtraque ad ventriculi gibbi sinistrum, et secunda satis obscurè ad ventriculi os procedit.
O In sinistrum fundi ventriculi, et superiorem omenti membranam: hac non mediocrem excrementi lienis portionem in ventriculum excerni putauerim.
P Numerose inter mesargi membranas distributa in intestina excurrit: ob hac ne, an 'a caua, hæmorrhoides sint? non ausim certo affirmare. Nam ex utraq; vena rami in eam partê ptinent, et etià maiores a porta: nec per portam melancholicum sanguinem expurgari, sortè alienum animaduertenti, apparebit. ⊱

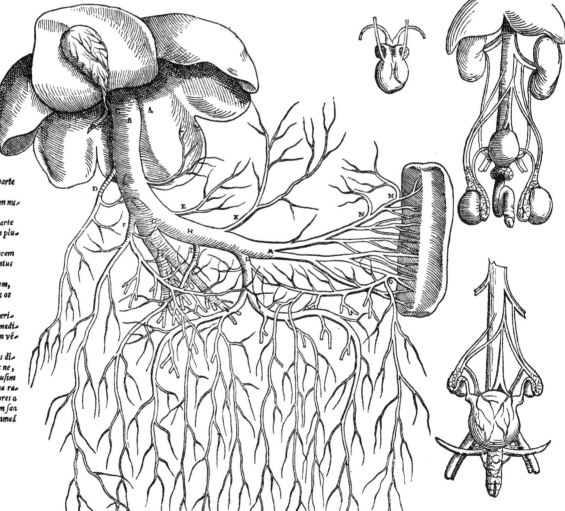

◊VENÆ CAVÆ, IECORARIÆ, ΚΟΙΛΗΣ, האורט HA⸗
NABVB DESCRIPTIO, QVA SANGVIS OMNIVM PARTIVM NVTRIMENTVM PER
VNIVERSVM·CORPVS·DIFFVNDITVR.

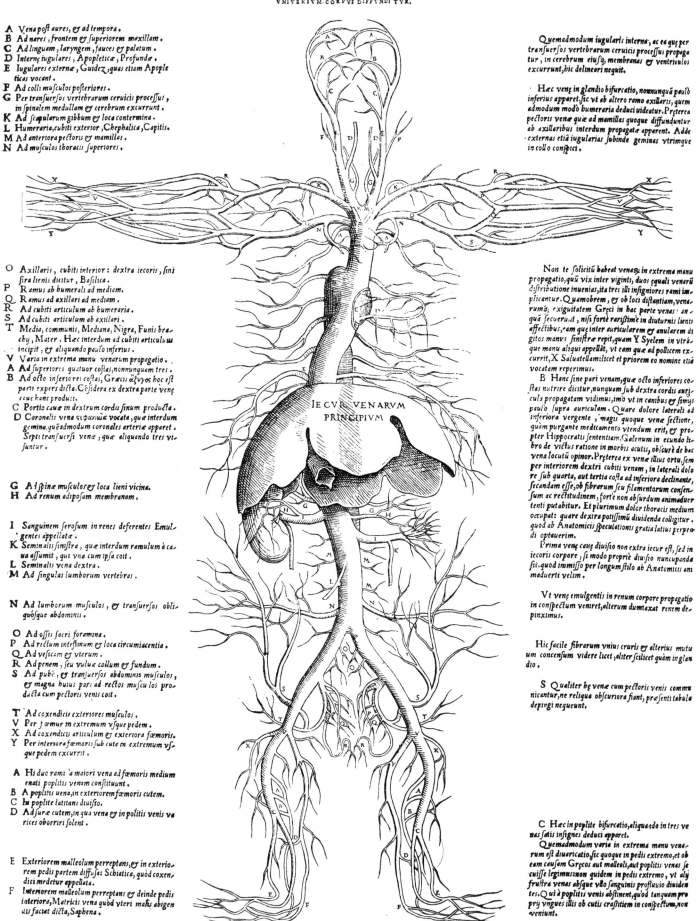

IECVR VENARVM PRINCIPIVM

A Vena poſt aures, ẽ ad tempora.
B Ad nares, frontem ẽ ſuperiorem maxillam.
C Ad linguam, laryngem, fauces ẽ palatum.
D Internæ iugulares, Apopleticæ, Profundæ.
E Iugulares externæ, Guidez, quas etiam Apople⸗
   ticas vocant.
F Ad colli muſculos poſteriores.
G Per tranſuerſos vertebrarum ceruicis proceſſus,
   in ſpinalem medullam ẽ cerebrum excurrunt.
K Ad ſcapulorum gibbum ẽ loca contermina.
L Humeraria, cubiti exterior, Chephalica, Capitis.
M Ad anteriora pectoris ẽ mamillas.
N Ad muſculos thoracis ſuperiores.

Quemadmodum iugularis internæ, ac ea quæ per
tranſuerſos vertebrarum ceruici proceſſus propaga⸗
tur, in cerebrum eiuſq̃ membranas ẽ ventriculos
excurrunt, hic delineari nequit.

· Hæc venæ in gladio bifurcatio, nonnunquam paulo
inferius apparet, ſic vt ab altero ramo axillaris, quem
admodum modo bumeraria deduci videatur. Præterea
pectoris venæ quæ ad mamillas quoque diffunduntur
ab axillaribus interdum propagatæ apparent. Adde
externas etiã iugularias ſubinde geminas vtrimque
in collo conſpici.

O Axillaris, cubiti interior: dextra iecoris, ſini⸗
   ſtra lienis dicitur, Baſilica.
P Ramus ab bumerali ad mediam.
Q Ramus ad axillari ad mediam.
R Ad cubiti articulum ab humeraria.
S T Ad cubiti articulum ab axillari.
T Media, communis, Mediana, Nigra, Funis bra⸗
   chy, Mater. Hæc interdum ad cubiti articuluⱰ
   incipit, ẽ aliquando paulo inferius.
V Varia in extrema manu venarum propagatio.
A Ad ſuperiores quatuor coſtas, nonnunquam tres.
B Ad octo inferiores coſtas, Græcis ἄζυγος hoc eſt
   paris expers dicta. Conſidera ex dextra parte venæ
   eſſe hanc producti.
C Portio cauæ in dextrum cordu ſinum producta.
D Coronalis vena ſεφανιαίa vocata, quæ interdum
   gemina, quæadmodum coronales arteriæ apparet.
   Septi tranſuerſi venæ, quæ aliquando tres vi⸗
   ſuntur.

Non te ſolicitũ habeat venæ in extrema manu
propagatio, quã vix inter viginti, duos æquali venarũ
diſtributione inuenias, ita tres illi inſigniores rami im⸗
plicantur. Quamobrem, ẽ ob loci diſtantiam, vena⸗
rumq̃ exiguitatem Græci in hac parte venas an⸗
quã ſecuerut, niſi forte rariſſime in diuturnis lienis
affectibus, eam quæ inter auricularem ẽ anularem di⸗
gitos manus ſiniſtræ repit, quam Y Syelem in vtrã⸗
que manu aliqui appellat, vt eam quæ ad pollicem ex⸗
currit, X Saluatellam licet et priorem eo nomine etiã
vocatam reperimus.

B Hanc ſine pari venam, quæ octo inferiores co⸗
ſtas nutrire dicitur, nunquam ſub dextra cordis auri⸗
cula propagatam vidimus, imo vt in canibus ẽ ſimys
paulo ſupra auriculam. Quare dolore laterali ad
inferiora vergente, magis quoque venæ ſectione,
quãm purgante medicamento vtendum erit, ẽ pro⸗
pter Hippocratis ſententiam, Galenum in ſecundo li⸗
bro de victus ratione in morbis acutis, obſcure de hac
vena locutũ opinor. Præterea ex venæ illius ortu, ſem
per interiorem dextri cubiti venam, in laterali dolo
re ſub quarta, aut tertia coſta ad inferiora declinante,
ſecandam eſſe, ob fibrarum ſeu filamentorum conſen⸗
ſum ac rectitudinem, forte non abſurdum animaduer
tenti putabitur. Et plurimum dolor thoracis medium
occupat, quare dextra potiſſimũ diuidenda colligitur.
quod ab Anatomicis ſpeculationis gratia latius perpen⸗
di optauerim.

G Ad ſpinæ muſculos ẽ loca lieni vicina.
H Ad renum adipoſam membranam.

Prima venæ cauæ diuiſio non extra iecur eſt, ſed in
iecoris corpore, ſi modo proprie diuiſio nuncupanda
ſit, quod immiſſo per longum ſtilo ab Anatomicis ani
maduerti velim.

I Sanguinem ſeroſum in renes deferentes Emul⸗
   gentes appellatæ.
K Seminalis ſiniſtra, quæ interdum ramulum à ca⸗
   ua aſſumit, qui vna cum ipſa coit.
L Seminalis vena dextra.
M Ad ſingulas lumborum vertebras.

Vt venæ emulgentis in renum corpore propagatio
in conſpectum veniret, alterum dumtaxat renem de⸗
pinximus.

N Ad lumborum muſculos, ẽ tranſuerſos obli⸗
   quóſque abdominis.

Hic facile fibrarum vnius cruris ẽ alterius mutu
um concenſum videre licet, aliter ſcilicet quãm in glan
dio.

O Ad oſſis ſacri foramina.
P Ad rectum inteſtinum ẽ loca circumiacentia.
Q Ad veſicam ẽ vterum.
R Ad penem, ſeu vuluæ collum ẽ fundum.
S Ad pubẽ, ẽ tranſuerſos abdominis muſculos,
   ẽ magna huius pars ad rectos muſculos pro⸗
   ducta cum pectoris venis coit.

S Qualiter bẽ venæ cum pectoris venis commu
nicantur, ne reliqua obſcuriora fiant, præſenti tabula
depingi nequeunt.

T Ad coxendicis exteriores muſculos.
V Per femur in extremum vſque pedem.
X Ad coxendicis articulum ẽ exteriora fœmoris.
Y Per interiora fœmoris ſub cute in extremum vſ⸗
   que pedem excurrit.

A Hi duæ rami à maiori vena ad fœmoris medium
   enati poplitis venam conſtituunt.
B A poplitis vena, in exteriorem fœmoris cutem.
C In poplite latitans diuiſio.
D Ad ſuræ cutem, in qua vena ẽ in politis venis va
   rices oboriri ſolent.

C Hæc in poplite bifurcatio, aliquando in tres ve
nas ſatis inſignes deduci apparet.
Quemadmodum varia in extrema manu vena⸗
rum eſt diuaricatio, ſic quoque in pedis extremo, et ob
eam cauſam Græcos aut malleoli, aut poplitis venas ſe
cuiſſe legimus, non quidem in pedis extremo, vt aly
fruſtra venas abſque vllo ſanguinis profluuio diuiden
tes. Qui à poplitis venis abſtinent, quod tanquam pro
prij vngues illis ob cutis craſſitiem in conſpectum, non
veniunt.

E Exteriorem malleolum perreptans, ẽ in exterio⸗
   rem pedis partem diffuſa: Schiatica, quòd coxen⸗
   dicis medetur appellata.
F Interiorem malleolum perreptans ẽ deinde pedis
   interiora, Matricis vena quòd vteri malis abigen
   dis faciat dicta, Saphena.

ALIQVI VENÆ CAVÆ RAMOS INSIGNIORES CENTVM ET SEXAGINTA OCTO POSVERVNT.

⚜ ARTERIA MAGNA, AOPTH, הנביב, HAORTI EX SI⸗
NISTRO CORDIS SINV ORIENS, ET VITALEM SPIRITVM TOTI CORPORI DEFERENS, NATV⸗
RALEMQVE CALOREM PER CONTRACTIONEM ET DILATATIONEM TEMPERANS.

A Plexus choriformis in cerebri anterioribus ventriculis ex arterijs & venis constitutum.
B Plexus reticularis ad cerebri basim, Rete mirabile, in quo vitalis spiritus ad animalē præparatur.
C Post aures, & ad tempora, & faciem arteriæ.
D Ad linguam, laryngem & fauces.
E Arteriæ καρωνδ́ες id est soporariæ, Apopleticæ, Subeticæ, סכרדים banirdamim.
F Ad transuersos vertebrarum ceruicis processus ad cerebrum vsque excurrentes.

G Ad pectoris os & mamillas, quæ cū illis quæ in rectis musculis sunt, communicantur.
H Ad humeri musculos & gibba scapularū.
I Ad supercostales musculos & mamillas.
K Sub axillari vena in brachium excurrit.
L Ad cubiti articulum vtrimque vna.
M In interna parte manus, & ramulus ad partem exteriorem pollicis.
N Ad superiores thoracis costas.
O Diuisio maxima, cuius maior ramus ad inferiorem corporis partem diffunditur, à quo mox in singulas costas propagines diuaricantur.
P Vena caua in dextrum cordis sinum aperta.
Q Arteria venalis in sinistrum sinum aerem ex pulmonibus deferens.
R Vena arterialis ex dextro sinu sanguinem pulmonibus communicans.
S Septi transuersi arteriæ satis insignes.
T In lienis sinū, pro visceris ratione maximæ.
V Ad iecoris cauum, & bilis vesicam.
X Ad ventriculum, & omentum.
Y In mesenterium par te superiori.
A Ad renes, Emulgentes dictæ, venis ipsis minores.
B Arteriæ seminales: vtrimque vna.
C Per mesenterium ad intestina vsque diffusa.
D Ad lumborum vertebras, musculos abdominis transuersos & obliquos.
F Ad foramina ossis sacri, quas nōnulli pro venis hæmorrhoidibus male demonstrare solēt.
G Ad vesicā, in viris ad penē, in mulieribus ad vuluæ fundum et collum.
H Arteriæ per quas fœtui spiritus in vtero cōmunicatur, quæ interdum in maiores truncos implantatæ conspiciuntur.
I Ad rectos abdominis musculos, cum pectoris arterijs cœeuntes, per quas vtero cum mamillis communio est.
K Ad coxendicis articulum, & fœmoris exteriorem regionem.
L In poplite bifurcatio in alto latens.
M Ad interiorem pedis partem latitans.
N Exteriorem pedu partem (licet profundè) perreptans, à quibus minimi rami in pedis superiorem partem excurrunt, ramulum tamen manifestum ad exteriora pollicis diffundunt.

Sinistram carotidem, aliquando ab ea quæ in sinistrum brachium fertur, deductam vidimus. sicut etiā ambas pectoris, ab ea quæ in dextram manum propagatur diuaricatas reperimus.

Arteriæ magnæ inæqualis diuisio, aliquando cordi vicinissimæ visitur, aliquando vero nonnihil à corde paululū remota, quemadmodum bis delineauimus.

Coronales arteriæ, in suo ortu demonstrari præsenti tabula nequeunt, latitāt enim post membranulas spiritum ex magna arteria in cor referri prohibentes.

Arteriæ quæ in iecur, lienem, ventriculum, omentum & mesenterium diffunduntur nonnunquam binas quemadmodum hic, sortuntur radices, interdū tres, & aliquando (licet in hominibus rarius) vnam. Verum semper propemodum ad hunc modum in transuersā ferri inuenimus.

Seminales arterias vtrasq́, postquam primū animaduerti, semper ab arteriæ magnæ corpore, aliter scilicet quam venas seminarias enatas inueni. licet etiam sinistram semel certissimè deesse repererim.

Hanc arteriam ad medium tibiæ vsque indiuisam ferri aliquando obseruauimus.

Hæc in extremo pede ac malleolo, sicuti etiam in extrema manu arteriarum distributio, subinde varieri consueuit. Verum quemadmodum sæpius nobis apparuit, hic detraximus.

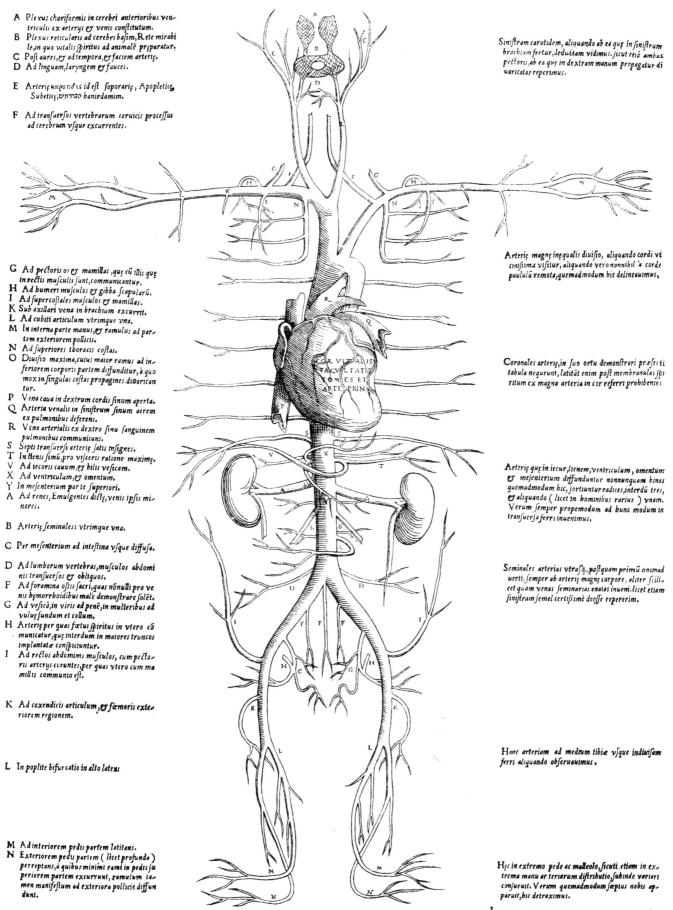

COR VITALIS
FACVLTATIS
FOMES ET
ARTE PRIN.

NOTATV DIGNAE ARTERIAE MAGNAE SOBOLES CENTVM ET QVADRAGINTA SEPTEM APPARENT.

❧HVMANI CORPORIS OSSA PARTE ANTERIORIEXPRESSA.

Foramina quæ in harum triũ chartarum delineatione conspici possunt ,sunt in temporum osse *auditorius meatus*: post mamillarem processum vnum, per quod interna iugularis in cerebrum *mergit*:in facie circa oculorum sedem quatuor, primum ad frontem, secundum ad nares, tertium *ad* maxillam superiorem, quartum ad temporalem musculũ:duo quoq; in maxilla inferiori. Et per *hæc* singula ramulus tertij paris neruorum excidit.

Ὀδόντες ,dentes, שיניים scinaim,plurimum triginta duo. τομεῖσ,incisorij,מחתבתים mechathchim, *octo* : κυνόδοντεσ,canini, ניבים calbym quatuor,μύλιται , molares , maxilleres, טוחנות thochnim *viginti*. omnes disparibus radicibus suos alueolos subeunt.

B Clauiculæ , κλᾶδεσ, claues,iugula, תרקוה tharkuha,Furculæ: vtrumꝗ; os literam.f.refert , figura inæquabili .

C Ἀκρώμιον , summus humerus,processus superior scapulæ , à Galeno in lib.de vsu par. κορακοει*δεσ* ad rostri corunii similitudinem nominans, עורב חזק alzegam charton,huius appendix *cuius* principio claues per arthrodian dearticulantur , proprie κατακλειδὰ quasi ad clauiculas *dicitur* , Rostrum porcinum .

D Processus scapulæ interior inferiórque ab anchoræ similitudine ἀνκυροειδὴσ dictus,ꝗ hunc *sæpe* dicitur ἀγκυροειδία ꝗ sigmoide Gale.vocauit. עוקץ עין hacatheph , Oculus scapulæ .

E Pectoris os , στερνον,החזה bechaseh , Cassos,septem constat ostibus ,sicut costæ quæ illi alli*gantur*,per vnionem potius,ꝗuàm per coarticulationem,parte inferiori iunctis:sid ab vtroꝗ, la*tere* lunatum est .

F Cartilago ξιφοειδὴσ ,ensiformis ,quo nomine totum os quoque dicitur ,הגרגרת alchangri,Ensi*fundis*,Malum granatum,Epiglottalis cartilago .

G Βραχίων , brachium,humerus Celso ꝗ Cæsari, זרוע Zeroach , Adiutorium brachy , Aseth: *hoc* tibiæ osse minus est .

H Sinus,humeri caput veluti in duo tubercula diuidens.

I Humeri orbita trochleis similis.

K Cubitus,πῆχυσ,הקנה kaneh, Asaid,quibus nominibus etiam tota hæc pars dicitur ,vlna. Fo*cile* matus , צלע אילון zenad elion . huius acutus processus ad brachiale ὠλεκρανον nominatur

L Radius, κερκίσ צמד תחתון zenad thachthon, Focile minus brachy .

N Brachiale, καρπὸσ, רשג reseg,Raseta,R ascha,ostibus disparibus octo ꝗ duplici ordine di*stinctis* constat,in superiori tribus, in inferiori quatuor:hęc simul figuram intrinsecus cauam, *ꝗ* extrinsecus gibbam constituunt:istorum cum Celso non incertus numerus est.

O Μεταχάρπιον , palma , pecten , המרק masrek , Postbrachiale ostibus quatuor Galeno , non *quinque*, vt alijs cũplurimis ,conformatum est .

P Δάκτυλοι , digiti, האצבעות esbaoth,singuli ex ternis ostibus conformantur,priori semper interno *dio* in subsequentis sinum subeunte .

Q Μύλη , ἐπιγονατίσ , patella , rotula genu, הארכובה magen harcubach ,scutum genu, Areʃ*fatu*: os rotundum breuis scuti instar .

R Ἀστράγαλοσ ,talus, הקרסול karsul,Balistæ os,Cauilla,Chabab,Alʃuchi:aliqui malleolum hodie *malè* vertunt .

S Nauiforme , σκαφοειδὴσ , nauiculare , זורקי zorki .

T πτέρνοσ, עקב reseg,Raseta pedis,quatuor ostibus constat,quorum maximum extrinsecus situm à *cuius* figura dicitur κυβοειδὴσ,tessera os , תארי thardy , Exagonon , Grandinosum ,Nerdi.Re*liqua* tria nominibus carent, sed κγλκοειδὴ nonnullis nominantur.Bis vidimus dextrum pedem *vno* abundare .

V Planta,planum , πεδίον , pecten המרק masrek , ostibus quinque constructum est , cui succedunt *pedis* digiti, X qui omnes ex ternis internodijs cõstant,magno tantũ excepto qui inter alios ex du*plici* osse constructus est .

Osticulum illud quod ad primum pollicis articulum apparet,vnum ex sesaminis ostibus est: *ꝗ* in illo duntaxat loco duo in vtroꝗ pede obseruauimus .

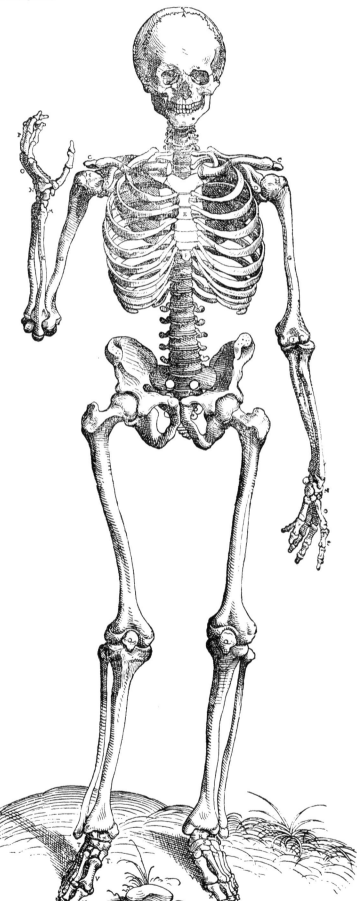

HVMANI CORPORIS OSSA NONNVLLI IN DVCENTA QVADRAGINTA OCTO, ALIQVI VERO, *in* alium numerum redigunt, ego excepto hyoide quod integrum fere ex sex osticulis per syncondrosim vnitis conformatur, ꝗ sesaminis ducenta ꝗ quadraginta sex putauerim sequentis tabellæ disticho comprehensa.

IV.

✠LATERALIS ΣΚΕΛΕΤΟΥ FIGVRÆ DESIG-
NATIO.

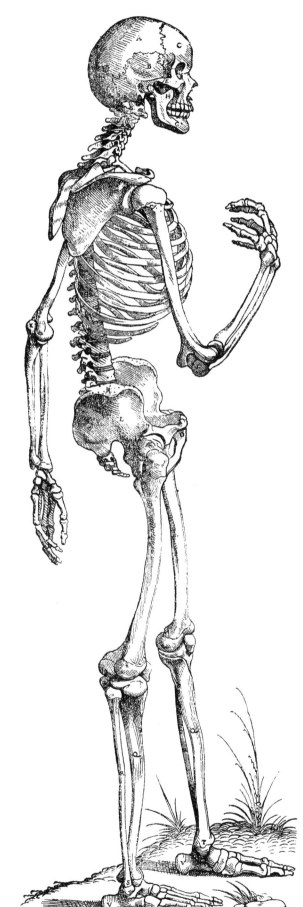

Offa κρανίου, caluaria, capitis offis, r.פיר קדרוש kadroth hamuach, Offę capitis, Afoan.

A Offa duo ϐρέγματοσ, κορυφῶσ, fincipitis, verticis, Parietalia. Locus hic apud Auicēnā in arabico et latino exemplari falfus eft.

B Offa duo ad vtráq; aurens κροταφῶν, temporum, Aurium: Singulorum ϐαλανοειδῶ id eft ecui, aut telo fimilis et mamillaris, et ad os iugale proceffus, partes exiftunt.

C Ὡς μετῶπου, frontis, נפצם מצ etzem hametzah, Coronale: interdū ob protenfam ad nafum vfque reclam futuram, geminum apparet, quod nonnulli in omnibus mulieribus effe falfò putarunt.

D Os vnum iniou, occipitis, מ oreph, cuneiforamen maximum ineft per quod fpinalis medulla excidit.

E Offa ξυγώματα, iugalia, aut ξυγοειδῆ, מ zog, Paris: vtrinque vnum ex duorum oftium conftantia proceffibus: quamobrem propria circunfcriptione carent.

F Os σφηνοειδῶσ, cuneiforme, bafilare, aliquando à multiplici forma πολύμορφον, נרח מוה mofchab hamoach, Colatory, Cauille, Hoc inter maxillę fuperioris offa decimū quintū numerari confueuit. Sunt enim fex quę ad radicem oculorum fubeunt: duo maxima malas ac molarium dentium alueolos continentia: narium duo : incifiorios dentes fufcipientia duo: ad fine palati circa narium foramina duo, et ante hęc omnia nuper dictum os cuneiforme. Nifi fortaffis octo, aut ex aliquorum Gręcorum fententia duodecim, hic offa enumerare meuis, prout fcilicet exiguas futuras, commiffuras et harmonias aut numeras, aut pręteris.

G Offa duo maxillę inferioris, parte anteriori per coalitum firmiffimè annexa: nec fat fcio, an malè cum Celfo in hominibus vnum dicere poffimus, nam uauis etiam decoctione feparari bauquaq; poffe obferuaui, et fi ipfa cultro dirimenda fit, nullibi difficilius quā in medio illam diuides.

H κορωνόν.

I Maxillę inferioris tuberculum et ceruix. hęc fola in omnibus animātibus pręterquā crocodilo mobilis dicitur.

K Duo cubiti proceffus, quorum pofteriorem ὠλεκρανὸν nominant. Hi in medio finum habent antiquę Gręcę literę Ϲ, aut noftrę C fimilem.

Coftæ, πλευραὶ צלעוש tzelaoth, viris et mulieribus viginti quatuor vtroq; latere duodecim: Ex ijs feptem cū metaphreni feu thoracis vertebris cunq; offe pecloris vtrinq; coeunt, quę verę et perfecté dicuntur. Cęterę quinque pofteriori parte fpinę duntaxat adueclūtur, ex quibus tres priores antica parte, fuis cartilagnibus, veris cohęrēt, alię duę inui cem dehifcunt. hę fpuriæ et nothæ et falfæ vocantur. fola autem duodecima vnica articulotione duodecimę vertebrę iungitur.

L Offa validiffima, quę offi facro cōmittuntur, ורבין צלגף hreua. Superne λαγόνων, ilium offa, הגרךש alzar gepha, Anchæ. ad fœmoris ingreffum ἰσχον, coxendicis, ךיר ezem haiarech, Pixis coxę, קרקר caph haiarech, Althauorat. O Parte anteriori qua tenuia ac forata mutuóque inter fe per fynchondrofim connexa funt, ἤβως, pubis, pectinis, altage, Penis dicuntur. Totū os, Celfo ex xę os, quemadmodum authori introduclo ry feu medici lgqcu, appellatura eft. Nonnulli fulsò putarunt hęc offa in viris ad pubem non effe per cartilaginem alligata.

Q Tibia, κνήμη, פצם אצמוש abt zmoth bafcek, qbus noftus tota hęc pars nolatur שוק cane godol, Canne maior, Focile cruris maius. Huius pars anterior excarnis et tenuis crea nominatur, huic tanummodo quoque fœmur annectitur: pręterea facilè tibię finus qui bus fœmoris capita recipit, apparent.

R Fibula, fura, os minus tibię, περόνη, שוק canne katon, Canne et arundo minor. Hoc os tibię craffitudine admodum cœdit, nec ita protenditur, vt genu ipfum contingat: verum fupra infráq; tibię per fynnartrofim coarticulatur. Tota hęc pars Celfo crus nominetur.

S.T Malleoli, σφυρά, קרצוב arcuboth, Clauiculę extremę tibię furęque proceffuum partes funt.

V Omnium pedis maximum os, κολκανέοσ, πτῆλα, calcis os, צרב aekef. huius pars pofterior tibię reclitudinem longè excedit.

DISTICHON OSSIVM NVMERVM COMPLECTENS.
Adde quater denis bis centum fenéque, habebis
Quàm fis multiplici conditus offe femel.

V.

♣ ΣΚΕΛΕΤΟΝ A TERGO DELINEATVM.

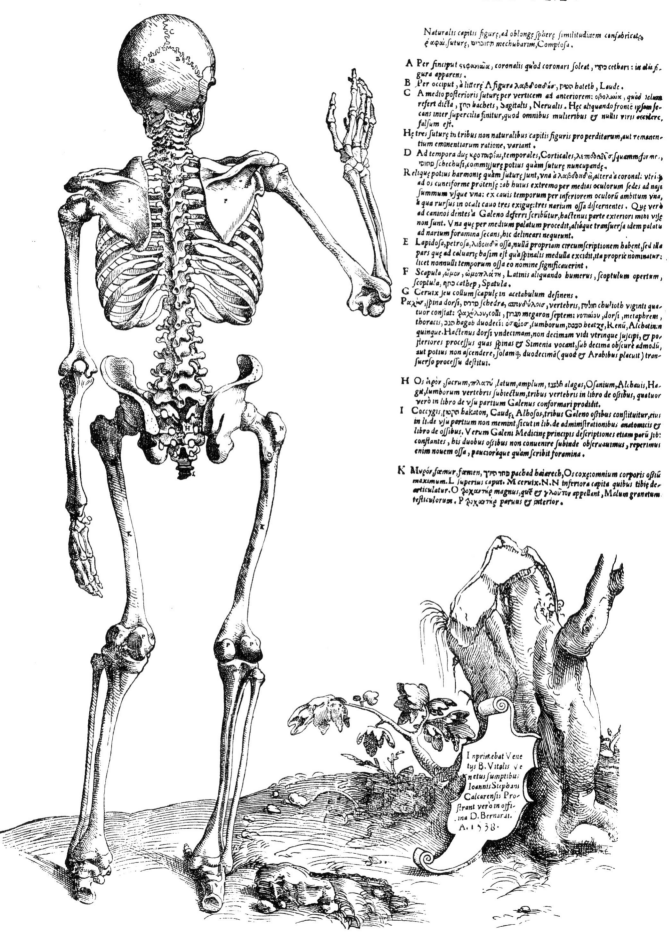

Naturalis capitis figura, ad oblongæ ſphæræ ſimilitudinem confabricatæ, ẽ ἀφαὶ ſuturæ, ὀιⁿοⁱⁿ mechubarim, Comploſa.

A Per ſinciput εγφανεία, coronalis quòd coronari ſoleat, פרו cethari : in alia figura apparet.

B Per occiput, à litteræ Λ figura λαβδοαδ̈ας, פרו haleth, Laude.

C A medio poſterioris ſuturæ per verticem ad anteriorem: οβολαία, quòd teluͫ refert dicta, פרו bacheth, Sagitalis, Neruialis. Hæc aliquando fronte ipſum ſecans inter ſupercilia finitur, quod omnibus mulieribus ꝗ nullis viris euenere, falſum eſt.

Hæ tres ſuturæ in tribus non naturalibus capitis figuris pro perditarum, aut remanentium eminentiarum ratione, variant.

D Ad tempora duæ κροταφίαι, temporales, Corticales, λιπδδαλ̈ ſ σ̈quammꝫ formæ, פרו ſchechuſh, commiſſuræ potius quàm ſuturæ nuncupandæ.

R eliquæ potius harmoniæ quàm ſuturæ ſunt, vna à λαβδοαδ̈ ε̈, altera à coronali: vtriꝗ ad os cuneiforme protenſæ: ab huius extremo per medias oculorum ſedes ad naſæ ſummum vſque vna: ex cauis temporum per inferiorem oculorum ambitum vna, à qua rurſus in oculi cauo tres exiguæ: tres narium oſſa diſcernentes. Quæ verò ad caninos dentes à Galeno deſerri ſcribûtur, hactenus parte exteriori mihi viſæ non ſunt. Vna quæ per medium palatum procedit, aliáque tranſuerſa idem palatũ ad narium foramina ſecans, hic delineari nequunt.

E Lapidoſa, petroſa, λιβδαδ̈ϖ oſſa, nullá propriam circumſcriptionem habent, ſed illa pars quæ ad caluariæ baſim eſt quàſpinalis medulla excidit, ita propriè nominatur: licet nonnulli temporum oſſa eo nomine ſignificauerint.

F Scapula, ὠμον, ὠμοπλάτη, Latinis aliquando humerus, ſcoptulum opertum, ſcoptula, פרו cathep, Spatula.

G Ceruix ſeu collum ſcapulæ in acetabulum deſinens.

P αχιϖ, ſpina dorſi, פרו ſchedra, αρυⁿφὐλϭιϭ, vertebris, פרⁿⁿ chuluoth viginti quatuor conſtat: ῥαχίλου, colli, פרⁿⁿ megaron ſeptem: νοτιαίον, dorſi, metaphrem, thoracis, פרⁿ hagab duodeci: ὀſφύοϭ, lumborum, פרⁿ beatzæ, Renũ, Alchatimⁿ quinque. Hactenus dorſi vndecimam, non decimam vidi vtrinque ſuſcipi, ꝗ poſteriores proceſſus quas ſpinas ꝗ Simenia vocant, ſub decima obſcurè admodũ, aut potius non aſcendere, ſolamꝗ duodecimã (quod ꝗ Arabibus placuit) tranſuerſo proceſſu deſtitui.

H Os ἱερὸν ſacrum, πλατύ, latum, amplum, פⁿ alagas, Oſanium, Alchauis, Hagel, lumborum vertebris ſubiectum, tribus vertebris in libro de oſſibus, quatuor verò in libro de vſu partium Galenus conformari prodidit.

I Coccygis, פⁿ bakaton, Caudæ, Alhoſos, tribus Galeno oſſibus conſtituitur, eius in li. de vſu partium non meminit ſicut in lib. de adminiſtrationibus anatomicis ꝗ libro de oſſibus. Verum Galeni Medicinæ principis deſcriptiones etiam perũ ſib: conſtantes, bis duobus oſſibus non conuenire ſubinde obſeruauimus, reperimus enim nouem oſſa, paucioréque quàm ſcribit foramina.

K Mⁿⲅόⲣ fæmur, fœmen, פⁿⁿ פⲁⲭ pachad baiarech, Os coxæ: omnium corporis oſtiũ maximum. L ſuperius caput. M ceruix. N. N inferiora capita quibus tibiæ de articulatur. O ῥοχατήρ magnus, quæ ꝗ γλϭϋτον appellant, Malum granatum teſticulorum. P ῥοχατήρ paruus ꝗ interior.

Inscription on scroll:
Imprimebat Venetys B. Vitalis Venetus ſumptibus: Ioannis Stephani Calcarenſis Proſtrant verò in offiⁱⁿⁿ D. Bernardi. A. 1538.